Residential Construction Performance Guidelines

for Professional Builders & Remodelers
Sixth Edition

NAHB Remodelers

Single-Family Builders Committee

Residential Construction Performance Guidelines for Professional Builders & Remodelers, Sixth Edition

BuilderBooks, a Service of the National Association of Home Builders

Patricia Potts	Senior Director
Design Central	Cover Design
Robert Brown, Jr.	Composition
Gerald M. Howard	Chief Executive Officer
John McGeary	SVP, Business Development & Brand Strategy
David Jaffe	Legal Review
Marcia Childs	Director, Business Management Department

Disclaimer

Printed in the United States of America

25 24 23   3 4 5

ISBN 978-0-86718-791-5
eISBN 978-0-86718-792-2
Library of Congress CIP information available on request.

For further information, please contact:
National Association of Home Builders
1201 15th Street, NW
Washingt on, DC 20005-2800
BuilderBooks.com

Contents

Acknowledgements

NAHB gratefully acknowledges the leadership and commitment of the groups that worked together to produce this publication and its predecessors:

Residential Construction Performance Guidelines, Sixth Edition:

Co-Chairs: Nicole Goolsby, AMB, AR, CGP, CAPS, CR, Red Ladder Residential, Inc.
　　　　　Robert Hanbury, CGR, Construction Solutions Consultant, LLC (Retired)

Review Group Leaders
John C. Cochenour, Lexington Fine Homes (Retired)
Tom Gotschall, CGP, Construction Arbitration Associations
Alan Hanbury, CAPS, CGR, GMR, House of Hanbury Builders, Inc. (Retired)
Peter Merrill, Construction Dispute Resolution Services, LLC
Michael Turner, CAPS, CGP, CGR, GMB, GMR, Installation Made Easy, Inc.
Kevin Woodward, Legacy Homes, Inc.

Work Group Participants
Kenneth Adams, CGB, CGP, CSP, Integrity Builders, Inc.
Dan Bawden, Legal Eagle Contractors Company
Mark Beliczky, Prohome International, LLC
Denis Bourbeau, Bourbeau Custom Homes, Inc.
Kenneth Boynton, Boynton Construction, Inc.
Sam Bradley, Sam Bradley Homes
Taylor Burton, Taylor Burton Company, Inc.
Jim Chapman, Jim Chapman Communities, Inc.
Kurt Clason, CAPS, CGR, Clason Remodeling Company
Robert Criner, CAPS, CGB, CGP, CGR, GMB, GMR, Criner Remodeling
Don Dabbert, Jr., Dabbert Custom Homes
Carol Eisenlohr, Legend Homes
Michael Elman, Construction Dispute Resolution Services
Bob Frost, Frost Construction Company
John Gill, Quality Builders Warranty Corporation
Jules Guidry, Highland Homes
John Hodgin, CAPS, John Hodgin Construction Co., Inc.
Joseph Irons, CAPS, CGP, CGR, GMB, GMR, Irons Brothers Construction, Inc.
David Jaffe, NAHB Staff
Ric Johnson, CAPS, Right at Home Technologies
Gary Kerns, Gary Kerns Homebuilders, LLC
Marcus Kuizenga, James Hardie Building Products
Roger Langford, Professional Warranty Service Corporation
James Leach, Centricity
Al LeCocq, LeCocq Construction Company
Michael LeCorgne, 2-10 Home Buyers Warranty
Mark Lewis, 2-10 Home Buyers Warranty
Mark Martin, Sandmark Custom Homes, Inc.

Chuck Miller, CAPS, CGB, CGP, CGR, CMP, CSP, GMB, GMR, Master CSP, MIRM, Chuck Miller Consulting, LLC
Kevin Miller, 2-10 Home Buyers Warranty
Josh Moore, James Hardie Building Products
Vince Napolitano, Napolitano Homes
Greg Paxton, CAPS, CGP, Mr. Handyman of Kanawha Valley
John Piazza, Sr., Piazza & Associates Consultants, Inc.
Russ Pies, Builders FirstSource
Don Pratt, CAPS, CECS of Michigan, LLC
Allen Ream, CGB, CGP, GMB, Montana Heritage Home Builders, Inc.
Dwight "Sonny" Richardson, Richardson Home Builders, Inc.
Jeff Schnellmann, Silliman CitySide Homes, LLC
Kevin Stablier, NorthLake Construction & Development, LLC
Sean Sullivan, CAPS, CGP, Living Stone Design + Build
Paul Thomas, 2-10 Home Buyers Warranty
Mary Uher, APA The Engineered Wood Association
Tim Van Curen, The Sherwin-Williams Company
Lora Vassar, CAPS, CGP, Arch Design
Billy Wards, Champion Builders, LLC
Shawn Woods, Ashlar Homes, LLC

This book would not exist today had it not been for the original outstanding commitment from many individuals and organizations who produced the first edition of *Residential Construction Performance Guidelines*:
Bill Asdal, Chair, NAHB Remodelers
Suzanne Grove, Chair, Single-Family Small Volume Builders Committee

The following organizations and groups contributed to all editions of the *Residential Construction Performance Guidelines*:
Home Innovation Research Labs (formerly NAHB Research Center)
NAHB Remodelers

NAHB Standing Committees
Business Management & Information Technology
Construction, Codes & Standards
Construction Liability, Risk Management & Building Materials
Construction Safety & Health
Custom Home Builders
Single-Family Builders

Other Organizations
Bonded Builders Home Warranty Association
Builders Association of Greater Indianapolis
Building Industry Association of Washington
Canadian Home Builders Association
Greater Atlanta Home Builders Association
Home Buyers Warranty Corporation
MacLellan Wolfson Associates
Metropolitan Builders Association of Greater Milwaukee
Michigan Association of Home Builders
Portland Cement Association
Professional Warranty Service Corporation
National Tile Contractors Association
National Wood Flooring Association
Quality Builders Warranty Corporation
Residential Warranty Company
Southern Nevada Home Builders Association
Texas Association of Builders
Texas Residential Construction Commission
Tile Council of America
2-10 Home Buyers Warranty, Inc.
U.S. Department of Housing and Urban Development

Introduction

Beyond building codes and local regulations, both contractors and their customers have long sought measurable benchmarks that deal with the expectations of performance in the goods and services provided by the residential construction industry. Although codes and regulations address matters of health, safety, and welfare, matching a consumer's expectations entails having objective criteria regarding performance. Upon this premise, the Residential Construction Performance Guidelines were developed and refined to offer achievable minimum levels of workmanship for the products delivered.

History of the *Residential Construction Performance Guidelines*

The core of these criteria was first established as a basis for coverage under the insured warranty program, initially offered some 30 years ago. More than 20 years ago, the NAHB Remodelers (previously the Remodelors Council) embraced a similar compilation of guidelines, which led to the published editions of *Quality Standards for the Professional Remodeler.* The joint effort of the NAHB Remodelers and the Single-Family Builders Committee (formerly the Single-Family Small Volume Builders Committee) culminated in the first edition of these guidelines in 1996. Many of the individual guidelines have remained time-honored measures of residential construction performance.

Indeed, the *Residential Construction Performance Guidelines* have become the accepted tool in many locations throughout the country for evaluating performance by parties under residential construction contracts when a dispute may lead to litigation or arbitration.

Sixth Edition Review

The performance guidelines in this manual were initially created and reviewed by more than 300 builders and remodelers, as well as representatives of specialty trade organizations. This exhaustive review resulted in a first edition that builders and remodelers used as a reference with confidence as they communicated with their customers. Subsequent editions, including this one, are substantially similar to the initial publication, although certain guidelines have been updated to make them easier for both contractors and consumers to understand.

A few guidelines were deleted and others were added to make the text more comprehensive and consistent with current building science advancements. NAHB members strive to continually refine these guidelines to promote greater mutual understanding of expectations between contractors and their customers. It is hoped that this will reduce the potential for disputes throughout the new construction and remodeling process.

Scope of the *Residential Construction Performance Guidelines*

These guidelines are a collection of minimum performance criteria and should be interpreted as such. Many contractors routinely build to tighter tolerances than those expressed here. Moreover, the guidelines are intended as a reference that should be interpreted with common sense. They should be applied only within the scope of the particular project built; they are not intended to answer all questions pertaining to construction workmanship that might arise in the course of a typical residential construction project. The guidelines selected for this manual deal with the issues that most frequently prompt questions for the contractor and the consumer.

The developers of these guidelines examined typical building techniques and quality measures based on a general overview of residential construction within the United States. Local variation in construction techniques, materials, requirements, and environmental conditions may render the application of some of these guidelines inappropriate either for evaluation or suggested remedy. In such cases, the parties should expressly provide in writing that some standard other than the related guideline will apply. Similarly, if a specific problem has prompted locally developed guidelines or practices, or changes in the prevailing building code, the parties may choose the local guideline or practice to take precedence over these guidelines.

These guidelines neither constitute nor are they intended to substitute for a warranty. However, parties may agree to incorporate the guidelines by referencing them within a warranty or within another construction contract provision. The guidelines are separate and distinct from any manufacturers' or fabricators' warranties that may apply to materials and products used in a project.

Contractors often refer to these guidelines in the Dispute Resolution sections of their contracts as the first step, prior to invoking any formal dispute resolution process such as mediation, arbitration, or litigation. Using these guidelines as the basic performance criteria has eliminated the need to invoke a formal dispute resolution process in many cases.

The use and application of the *Residential Construction Performance Guidelines* is strictly optional and completely at the discretion of individual users. If they are used, they refer only to contractor-installed materials and services and not to any elements the consumer has contracted for through a third party, including other trade contractors.

Scope of Responsibilities

Typically, numerous parties are involved in a residential construction project, whether it entails building a new home or remodeling an existing one. Each of these parties has specific responsibilities to fulfill. The contract documents should provide a clear statement of the agreement between the contractor and the consumer. In addition to the specific provisions of any contract, the following general responsibilities should be noted:

Contractor: For the purposes of this manual, the contractor is the entity named in the contract that has primary responsibility for completing the project. The contractor often employs others to assist in the project. In most cases, the contractor is responsible for all work assigned in the contract regardless of who performs the work. If the contractor is acting in a special role (for instance, as a construction manager), or the consumer selects others to work on the project who are outside the contractor's control, then the responsibility for evaluation and remedy of potential problems will fall to those other parties.

Consumer: The consumer is the buyer of the product or service named in the contract. As such, the consumer is responsible for carefully reviewing the contract to ensure it accurately represents the expectations for the final product. Once the consumer accepts the project and moves into the home or occupies the newly reno-vated space, then the consumer is responsible for routine maintenance and upkeep. Homes require a certain amount of care and maintenance, which are the consumer's responsibility. Consumers should note that in some of the guidelines contained herein, the contractor is not obligated to make repairs to items that fall within the consumer's maintenance responsibilities.

Manufacturer: Manufacturers warrant many residential construction components that may fall outside the scope of the contractor's responsibilities, such as kitchen appli-ances, furnaces, air conditioners, and lighting and plumbing fixtures. Certain types of siding, roofing, or flooring also may be covered by a manufacturer's warranty. If there is a warranty question with one of these components, the consumer should be aware that the contractor may not be responsible for the product's performance after installation. If a problem occurs, the consumer often will need to contact the manufacturer or fabricator directly to have the problem evaluated and, if necessary, rectified, unless otherwise specified in a contract. The contractor's responsibilities may end once the contractor provides the appropriate information on how to contact the manufacturer or fabricator, unless otherwise specified in the contract.

Remodeling Projects

Remodeling, the process of expanding or enhancing an existing structure, presents inherent difficulties in melding the new and old into a home or room that meets the consumer's needs and is aesthetically pleasing. Some circumstances call for the suspension of some or all the guidelines to successfully complete a remodeling project. These circumstances include, but are not limited to, the meeting of old, out-of-plumb or out-of-level structures with new structures; the appearance of new materials near weathered, existing materials; aging plumbing and mechanical systems and the practical considerations for new projects to work within the limitations of existing buildings.

Because of the unique challenges of joining new and old construction, a remodeling contractor may build some or all of the project outside the scope of these guidelines to achieve the contract objectives. The contractor may note an existing condition with the consumer before construction. It is also normal for a contractor, during the course of construction, to discover and accommodate conditions in the old structure that require different solutions from those suggested in these guidelines. In these circumstances, the governing factor is meeting the needs of the consumer as outlined in the contract.

How to Use This Manual

This manual is divided into chapters organized according to the usual sequence of events in the construction process. Nearly every chapter contains sections within it, and some chapters also have more specific subsections. Each chapter contains individual construction performance guidelines.

The guidelines are numbered as follows:

Chapter Number–Section Number–Guideline Number

Each construction performance guideline has three parts, as follows:

Observation: A description of a particular construction condition.

Performance Guideline: The specific criterion for acceptable workmanship.

Corrective Measure: A description of the work required to meet the performance guideline and/or the consumer's maintenance responsibility.

Some guidelines also include the following elements:

Remodeling Specific Performance Guideline (listed as applicable): The specific criterion for acceptable workmanship for remodeling.

Discussion: An explanation of unique factors pertinent to the observation, performance guideline, or corrective measure.

General Instructions

Many locales require construction work to comply with the prevailing building code. If a conflict arises between these guidelines and the prevailing building code, as a matter of law, the code requirements may take precedence over these guidelines.

These performance guidelines apply only to work specified in the contract documents for the project. They do not apply to designs, plans, materials, or workmanship that is supplied by the consumer or is outside the scope of the project. They are also designed to apply only to the part of the job addressed in each guideline.

In order to obtain a specific result, many plans or specifications utilize styles, materials, or specific workmanship details that are designed to be outside of the specifications herein, to obtain a specific result. In order to avoid conflict, the contractor and consumer should specify in writing that these specific items are excluded from the performance guidelines.

Definition of Terms

The following terms are crucial to understanding the *Residential Construction Performance Guidelines, Sixth Edition:*

Substantial completion of the project. The point at which the areas in the home are functional for their intended use as stated in the contract. Substantial completion for a remodeling project is when the consumer occupies or uses the space for its intended purpose. The contract should include a specific definition of completion.

Warranty period. A 12-month period unless otherwise specified by the contract or prevailing state or local law.

Manufacturer's warranty. The warranty provided by the manufacturer of a product that has been incorporated into a newly-constructed home or a remodeling project.

Prevailing building code. The building code that has been adopted by the state, county, city, or other applicable local governing authority. These codes vary greatly and require an understanding of the codes specifically applicable to each individual project.

Other selected terms are italicized on first reference and defined in the glossary.

Incorporating the Guidelines into a Warranty or Dispute Resolution Program

The warranty, like the contract, should clearly express the intent of the parties. The limited warranty describes the issues the contractor will be responsible for after substantial completion of the project and specifies the time period during which the warranty is in force. Moreover, if a contractor warrants workmanship and materials in a warranty, the contractor should provide a clear definition of compliance with the terms of the warranty. Failure to do so may result in having an arbitrator or a court decide for them. Accordingly, consideration should be given to incorporating the *Residential Construction Performance Guidelines* by reference into the contract.

A contract may validly include provisions of a document that is not physically part of the contract itself – this is referred to as "incorporation by reference." Incorporation by reference is a common tool in the drafting of construction contracts. Matters incorporated into a contract by reference are as much part of the agreement as if they had been set out in the contract word-for-word. Consideration should be given to incorporating by reference into the contract the *Residential Construction Performance Guidelines.* Please refer to the Appendix to see examples of suggested contract language to incorporate the RCPG into your contract, warranty, or related document.

To ensure that the consumer agrees with the specific performance guidelines stated herein, the contractor should review the specific guidelines and the procedures recommended with the consumer before entering into a contract. Providing the consumer with a copy of the guidelines at contract signing is highly recommended. Reviewing the performance guidelines again at closing or at the walk-through inspection is also recommended.

If there are guidelines within this publication that the contractor or consumer does not want to use, they should be specifically excluded in writing from all warranty or contract documents applicable to the project. Likewise, if there are issues that are not addressed in the guidelines, then by written agreement the contractor and consumer should refer to those issues in the warranty and/or contract documents.

Other Uses for the Guidelines

The *Residential Construction Performance Guidelines* can promote a better understanding of the home construction process among consumers, inspectors, and public officials. Following are some suggestions for building awareness of these guidelines:

- Make the guidelines available to consumers simultaneous with or prior to entering into a contract to help them understand the construction process. Whether or not your contract refers to the guidelines, the contractor and consumer should acknowledge in writing that they agree to specific performance tolerances.

- Avoid disputes by referring to this objective set of guidelines with third-party credibility.

- Share the guidelines with mediators, arbitrators, and judges to help them understand the acceptable performance criteria.

- Show the guidelines to building code officials so they can distinguish performance guidelines from code compliance issues.

- Make the guidelines available to trade contractors whose profession is integral to the construction process.

- Ensure agreements with trade contractors include a guarantee from them that their work will comply with the guidelines.

- Take the guidelines to city, county, and state officials and urge them to consider adopting the guidelines as their accepted criteria.

- Make the guidelines available to private, third-party home inspectors, and their trade associations.

Taking Measurements

You can quickly assess whether certain ridges, cracks, gaps, lippage, or variations in plumbness or levelness are within the *Residential Construction Performance Guidelines* recommended tolerances. Tolerances in most of these areas are less than 1 inch. The edges of U.S. coins can be used to approximate measurements of variation as follows:

Dime = approximately 1/32 inch
Quarter = approximately 1/16 inch

Comments Welcomed

NAHB encourages readers to provide comments and suggestions regarding their experiences using the *Residential Construction Performance Guidelines, Sixth Edition,* including their own methods or tools for determining whether a project complies with the guidelines. Submit your comments in writing, with the subject line "Residential Construction Performance Guidelines, Sixth Edition" to the NAHB Business Management Department, 1201 15th Street NW, Washington, DC 20005. Comments will be considered in preparing future editions of this book.

Site Work

1–1–1 **Observation:** The ground has settled around the *foundation,* over utility trenches, or in other areas.

Performance Guideline: The ground will slope away from the foundation at least 6 inches within the first 10 feet of the building. Settled ground, outside of the building's 10-foot line, that is over utility trenches, or in other filled areas shall not interfere with water drainage away from the home.

Corrective Measure: If the contractor provided final grading, then one time only during the *warranty period,* the contractor will fill areas that settle more than 6 inches or that affect proper drainage. Any removal, repair, replacement or modification of landscaping, including but not limited to grass, sod, plantings, or lawn irrigation systems, etc., by the consumer shall negate any obligation on behalf of the builder. The contractor will make a good faith effort to preserve plantings, but it is the consumer's responsibility to replace shrubs, grass, other landscaping, pavement, sidewalks, or other improvements affected by placement of such fill.

1–1–2 **Observation:** The property does not properly drain.

Performance Guideline: To ensure proper drainage within 10 feet around the home, the contractor will establish necessary grades and swales within the property if the work is included in the contract. Standing water should not remain for extended periods (generally no more than 24 hours), within 10 feet of the home after a rain, except in swales that drain other areas or in areas that receive discharge from sump pumps, down spout drain lines, foundation or crawl space drains, etc. In these areas a longer period can be anticipated (generally no more than 48 hours). Water may stand longer during periods of heavy rains, especially when heavy rains occur on successive days. Removal, repair, replacement, or modification of landscaping, including but not limited to grass, sod, plantings, or lawn irrigation systems, etc., by the consumer shall negate any obligation on behalf of the builder. No grading determination will be made while frost or snow is on the ground or while the ground is saturated.

Remodeling Specific Guideline: Contractor is not responsible for problems caused by prior construction, preexisting conditions, changes in code requirements (prior codes caused problems later, for example), or existing materials and surfaces.

Corrective Measure: If grading is part of the contract, the contractor is responsible for initially establishing the proper grades and swales.

Discussion: Grass and other landscaping are integral components of the storm water management practice needed to minimize erosion from the property. It is the consumer's responsibility to maintain grass and other landscaping to help ensure the property drainage system functions properly. The consumer is responsible for maintaining such grades and swales once the contractor has properly established them. Irrigation of new plants is especially vital to their health and survival. In some seasons, daily water may be necessary to be applied by consumers until plants have established themselves into new soil, their new home.

1–1–3   **Observation:** The property has soil erosion.

    Performance Guideline: The contractor is not responsible for soil erosion.

    Corrective Measure: No corrective action is required by the contractor.

1–1–4   **Observation:** Water from a nearby or adjacent property flows onto the consumer's property.

    Performance Guideline: The contractor is responsible for providing a reasonable means of draining water from rain, melting snow, or ice on the property, but the contractor is not responsible for water flowing from a nearby or adjacent property.

    Corrective Measure: No corrective action is required by the contractor.

1–1–5   **Observation:** Existing trees, shrubs, or other vegetation were damaged during construction.

    Performance Guideline: The contractor will use their best judgment in making a reasonable and cost-effective effort to preserve existing landscaping as predetermined by the contractor and consumer, but the survival of existing landscaping cannot be guaranteed.

    Corrective Measure: No corrective action is required by the contractor.

1–1–6   **Observation:** Grass, grade or landscape areas are disturbed or damaged on the property as a result of work performed by the contractor in conjunction with a correction of a deficiency.

Performance Guideline: Grass, grade or landscape areas disturbed or damaged will be restored.

Corrective Measure: Contractor will repair disturbed or damaged areas affected during construction.

Discussion: Damage to grass, grade or landscaping caused by a third party, e.g., a utility provider, is not the responsibility of the contractor. Replacement material shall be of the same size and variety as the dead or damaged material.

Foundation

General

2–1–1   **Observation:** The foundation is out of square.

Performance Guideline: As measured at the top of the foundation wall, the diagonal of a triangle with sides of 12 feet and 16 feet will be no more than 1 inch more or less than 20 feet.

> **Remodeling Specific Guideline:** A contractor and consumer may agree to build an addition out of square in order to keep a new exterior wall in line with an existing wall of an out-of-square home.

Corrective Measure: The contractor will make necessary modifications to the foundation not complying with the performance guideline for squareness to provide a satisfactory appearance. The contractor may square the first-floor deck or walls by cantilevering over the foundation or locating the deck or walls inset from the outside face of the foundation.

Discussion: Squareness is primarily an aesthetic consideration. The corrective measure emphasizes the primarily aesthetic nature of squareness and makes the criterion for correction satisfactory appearance. This allows the contractor to make either a structural change or some cosmetic modification as is most appropriate. There are many instances in which the squareness of a foundation is inconsequential because subsequent construction provides an opportunity to make corrections.

2–1–2   **Observation:** The foundation is not level.

Performance Guideline: This guideline applies only when the levelness of the foundation adversely impacts subsequent construction. As measured at the top of the foundation wall, no point will be more than ½ inch higher or lower than any point within 20 feet.

> **Remodeling Specific Guideline:** The contractor and the consumer may agree to build an addition out of level to keep the floor of an addition on the same plane, and the roof ridge on the same line, as those of an existing, out-of-level structure.

Corrective Measure: The contractor will make necessary modifications to any part of the foundation or to subsequent construction to meet the performance guideline for levelness. This can be affected by leveling the sills with *shims, mortar,* appropriate fillers, or other methods.

Discussion: There are many instances in which the levelness of a foundation is not of consequence because subsequent construction provides an opportunity to make corrections.

2–1–3   **Observation:** There is a crack in a concrete footing.

Performance Guideline: Cracks greater than ¼ inch in width are considered excessive.

Corrective Measure: The contractor will repair any cracks in excess of the performance guideline, using a material designed to fill cracks in concrete.

Concrete Slabs

2–2–1   **Observation:** A concrete *slab* within the structure has separated or moved at control (expansion and contraction) joints.

Performance Guideline: Concrete slabs within the structure are designed to move at *control joints.*

Corrective Measure: Because this is a normal occurrence, no corrective action is required by the contractor.

Discussion: Control joints are placed in concrete for the very purpose of encouraging separation or cracking to take place at the joints instead of in random locations.

2–2–2   **Observation:** Efflorescence is present on the surface of the basement floor.

Performance Guideline: If the efflorescence is caused solely by basement water leakage (actual flow and accumulation), the contractor will eliminate the leaks into the structure.

Corrective Measure: The contractor will repair to meet the performance guideline.

Discussion: Efflorescence is a typical condition caused by moisture reacting with the soluble salts in concrete and forming harmless carbonate compounds. It is evidenced by the presence of a white film or powder on the surface of the concrete. It is a particularly common occurrence where masonry or concrete are in contact with high moisture levels as may be found in basements or crawl spaces.

2–2–3   **Observation:** The concrete floor or slab is uneven.

Performance Guideline: Except where the floor or portion of the floor has been designed for specific drainage purposes, concrete floors in living areas should not have pits, depressions, or areas of unevenness exceeding ⅜ inch in 32 inches.

> **Remodeling Specific Guideline:** Existing concrete floors or slabs may be uneven. In these situations, no corrective action is required by the contractor.

Corrective Measure: The contractor will correct or repair the floor to meet the performance guideline.

Discussion: A repair can be accomplished by leveling the surface with a material designed to repair uneven concrete.

2–2–4   **Observation:** The concrete floor slab is cracked.

Performance Guideline: Minor cracks in concrete floor slabs are normal. Cracks exceeding 3/16 inch in width will be repaired.

Corrective Measure: Using a material designed to fill cracks in concrete, the contractor will repair horizontal cracks that do not meet the performance guideline.

Discussion: If the cracks are observed at the control joints no corrective action is required by the contractor. Control joints are placed in concrete for the very purpose of encouraging separation or cracking to take place at the joints instead of in random locations.

2–2–5   **Observation:** The concrete floor has a crack that has a vertical displacement.

Performance Guideline: Minor cracks in concrete floor slabs are normal. Cracks exceeding 3/16 inch in vertical displacement will be repaired.

Corrective Measure: Using a material designed to fill cracks in concrete, the contractor will repair vertical cracks that do not meet the performance guideline. Any crack that creates a possible trip hazard will be ground down.

2–2–6 **Observation:** Interior concrete is *pitting* or *spalling*. Pitting is evidenced by concrete that has chipped. Spalling is evidenced by concrete that has flaked or peeled from the outer surface.

Performance Guideline: Interior concrete surfaces should not pit or spall unless the deterioration is caused by factors outside of the contractor's control.

Corrective Measure: The contractor will repair concrete surfaces using materials designed for this purpose.

2–2–7 **Observation:** The interior concrete slab, i.e. basement or garage slab, has a loose, sandy surface, sometimes referred to as "dusting."

Performance Guideline: The surface should not be so sandy that it creates a problem for flooring being installed by the owner after the contractor has completed the project.

Corrective Measure: The surface will be repaired to be suitable for the finish flooring that the contractor had reason to anticipate would be applied.

Concrete Block Basement and Crawl Space Walls

2–3–1 **Observation:** A concrete block basement or crawl space wall is cracked.

Performance Guideline: Cracks in concrete block basement or crawl space walls should not exceed ¼ inch in width.

Corrective Measure: The contractor will repair cracks to meet the performance guideline using a material designed to fill cracks in concrete.

Discussion: *Shrinkage cracks* are common in concrete block masonry and should be expected in crawl space and basement walls. Cracks may be vertical, diagonal, horizontal, or in stepped-in masonry joints.

2–3–2   **Observation:** A concrete block basement wall is out of plumb.

Performance Guideline: Concrete block walls should not be out of plumb greater than 1 inch in 8 feet when measured from the base to the top of the wall.

> **Remodeling Specific Guideline:** If tying into an existing foundation that is out of plumb, the contractor and consumer will review the existing conditions and scope of work. The contractor will use his or her best judgment in making a reasonable and cost-effective effort to meet the performance guideline while complying with the existing building code.

Corrective Measure: The contractor will repair any deficiencies in excess of the performance guideline unless the wall is to remain unfinished according to the contract, and the wall meets building code requirements as evidenced by passed inspections, in which case no corrective action is required by the contractor.

2–3–3   **Observation:** A concrete block basement wall is bowed.

Performance Guideline: Concrete block walls should not bow in excess of 1 inch in 8 feet.

Corrective Measure: The contractor will repair any deficiencies that do not meet the performance guideline unless the wall is to remain unfinished according to the contract, and the wall meets building code requirements, in which case no corrective action is required.

2–3–4   **Observation:** Efflorescence is present on the surface of the basement or crawl space concrete block wall.

Performance Guideline: If the efflorescence is caused solely by water leakage (actual flow and accumulation), the contractor will eliminate the leak into the structure.

Corrective Measure: The contractor will repair to meet the performance guideline.

Discussion: Efflorescence is a typical condition caused by moisture reacting with the soluble salts in concrete and forming harmless carbonate compounds. It is evidenced by the presence of a white film or powder on the surface of the concrete. It is a particularly common occurrence where masonry or concrete are in contact with high moisture levels as may be found in basements or crawl spaces.

Poured Concrete and Concrete Panel Basement, and Crawl Space Walls

2–4–1   **Observation:** A concrete basement wall is out of plumb.

      Performance Guideline: Finished concrete walls should not be out of plumb greater than 1 inch in 8 feet when measured vertically.

      Remodeling Specific Guideline: If tying into an existing foundation that is out of plumb, the contractor and consumer will review the existing conditions and scope of work. The contractor will use his or her best judgment in making a reasonable and cost-effective effort to meet the performance guideline while complying with the existing building code.

      Corrective Measure: The contractor will repair any deficiencies that do not meet the performance guideline. If the wall is to remain unfinished according to contract and the wall meets building codes, no corrective action is required by the contractor.

2–4–2   **Observation:** An exposed concrete wall has pits, surface voids, or similar imperfections in it.

      Performance Guideline: Surface imperfections larger than 1 inch in diameter or 1 inch in depth are considered excessive.

      Corrective Measure: The contractor will repair imperfections, which do not meet the performance guideline, using a material designed to fill holes in concrete.

      Discussion: Pits, surface voids, and similar imperfections are called "air surface voids" and are caused by air trapped between the concrete and concrete form interface. Air surface voids are not structurally significant. The technical term for larger voids is honeycomb. These must be dealt with in accordance with this guideline. The repaired area is unlikely to match the color or texture of the surrounding concrete.

2–4–3   **Observation:** A concrete basement wall is bowed.

      Performance Guideline: Concrete walls should not bow in excess of 1 inch in 8 feet when measured from the base to the top of the wall.

      Corrective Measure: The contractor will repair any deficiencies that do not meet the performance guideline. If the wall is to remain unfinished according to contract and the wall meets building codes, no corrective action is required by the contractor.

2–4–4   **Observation:** A concrete basement or crawl space wall is cracked.

Performance Guideline: Cracks in concrete walls should not exceed ¼ inch in width.

Corrective Measure: Using a material designed to fill cracks in concrete, the contractor will repair any cracks to meet the performance guideline.

Discussion: Shrinkage cracks and other cracks are common and are inherent in the drying process of concrete walls. They should be expected in these walls due to the nature of concrete. The only cracks considered under warranty claims are cracks that permit water penetration or horizontal cracks that cause a bow in the wall.

2–4–5   **Observation:** A *cold joint* is visible on exposed poured concrete foundation walls.

Performance Guideline: A cold joint is a visible joint indicating where the pour terminated and continued. Cold joints are normal and should be expected to be visible. Cold joints should not be an actual separation or a crack that exceeds ¼ inch in width.

Corrective Measure: Using a material designed to fill cracks in concrete, the contractor will repair any cold joint to meet the performance guideline.

2–4–6   **Observation:** A *joint* is visible on exposed panel concrete foundation walls.

Performance Guideline: A visible joint indicating where the panels connect are normal and should be expected to be visible. Joints should not be an actual separation.

Corrective Measure: Using a material designed to fill cracks in concrete, the contractor will repair any panel joint to meet the performance guideline.

2–4–7   **Observation:** Efflorescence is present on the surface of the concrete basement wall.

Performance Guideline: If the efflorescence is caused by basement water leakage (actual flow and accumulation), the contractor will eliminate the leak into the structure.

Corrective Measure: The contractor will repair to meet the performance guideline.

Discussion: Efflorescence is a typical condition caused by moisture reacting with the soluble salts in concrete and forming harmless carbonate compounds. It is evidenced by the presence of a white film or powder on the surface of the concrete. It is a particularly common occurrence where masonry or concrete are in contact with high moisture levels as may be found in basements or crawl spaces.

Moisture and Water Penetration

Basement Walls and Floor

2–5–1   **Observation:** Dampness is evident on basement walls or the floor.

Performance Guideline: The contractor is not responsible for dampness caused by *condensation* of water vapor on cool walls and floors. Dampness caused by moisture intrusion should be addressed by the contractor.

Corrective Measure: The contractor will repair to meet the performance guideline unless the consumer's action caused the dampness.

Discussion: Excessive dampness caused by consumer action, such as changing the grade around the home or irrigation systems, is not the contractor's responsibility.

2–5–2   **Observation:** Water has accumulated in the basement.

Performance Guideline: Water should not accumulate in the basement.

Remodeling Specific Guideline: The contractor is not responsible for problems caused by prior construction, preexisting conditions, changes in code requirements (prior codes caused problems later), or existing materials and surfaces and installation techniques and technology.

Corrective Measure: The contractor will take such actions as are necessary to prevent water from accumulating in the basement unless consumer action caused the accumulation.

Discussion: Water accumulation caused by consumer action, such as changing the grade around the home or irrigation systems, is not the contractor's responsibility.

Crawl Spaces

2–5–3   **Observation:** Water accumulates in a vented crawl space.

Performance Guideline: Crawl spaces should be graded and proper exterior foundation drains installed as required by the prevailing building codes to prevent water from accumulating.

Remodeling Specific Guideline: The contractor is not responsible for problems caused by prior construction, preexisting conditions, changes in code requirements (prior codes caused problems later), or existing materials and surfaces and installation techniques and technology.

Corrective Measure: The contractor will take corrective measures to meet the performance guideline.

2–5–4   **Observation:** Condensation is evident on the vented crawl space surfaces.

Performance Guideline: The contractor will install the ventilation and vapor barrier required by the prevailing building code.

Corrective Measure: The contractor will take corrective actions to meet the performance guideline. If the crawl space is ventilated as required by applicable building codes, then no corrective action is required by the contractor. Further reduction of condensation is a consumer maintenance responsibility.

Discussion: Temporary conditions that cause condensation that cannot be eliminated by ventilation and a vapor barrier may include:

- Night air gradually cools the interior surfaces of the crawl space. In the morning, moisture picked up by sun-warmed air migrates into the crawl space and condenses on cool surfaces.

- At night, outside air may rapidly cool foundation walls and provide a cool surface on which moisture may condense.

- If the home is left unheated in the winter, floors and walls may provide cold surfaces on which moisture in the warmer crawl space air may condense.

- Moisture inside a heated home may reach the dew point within floor insulation or on the colder bottom surface of the vapor barrier. Moisture on or under the poly vapor barrier may result from condensation or hydrostatic pressure. This is a normal condition.

- The consumer can reduce condensation, if necessary, by enclosing the crawl space and dehumidifying *(closed crawl)* or by enclosing and intentionally heating and cooling the crawl space *(conditioned crawl)*.

2–5–5   **Observation:** Sealed crawl/closed/conditioned has moisture or standing water.

Performance Guideline: Sealed crawl spaces should have adequate methods to drain the possible sources for ground moisture entering the space as required by applicable building codes.

Remodeling Specific Guideline: The contractor is not responsible for problems caused by prior construction, preexisting conditions, changes in code requirements (prior codes caused problems later), or existing materials and surfaces and installation techniques and technology.

Corrective Measure: The contractor will take corrective measures to meet the performance guideline.

Structural Columns, Posts, or Piers

2–6–1   **Observation:** An exposed wood column is bowed or is out of plumb.

Performance Guideline: Exposed wood columns should not bow or be out of plumb more than ¾ inch in 8 feet at substantial completion of the project.

Corrective Measure: The contractor will repair any deficiencies that do not meet the performance guideline.

Discussion: Wood columns may become distorted as part of the drying process. Bows and other imperfections that develop after installation cannot be prevented or controlled by the contractor.

2–6–2   **Observation:** An exposed concrete column is installed bowed or out of plumb.

Performance Guideline: Exposed concrete columns should not be installed with a bow more than 1 inch in 8 feet. They should not be installed out of plumb more than 1 inch in 8 feet.

Corrective Measure: The contractor will repair any deficiencies that do not meet the performance guideline.

2–6–3   **Observation:** A masonry column or pier is out of plumb.

Performance Guideline: Masonry columns or piers should not be constructed out of plumb more than 1 inch in 8 feet.

Corrective Measure: The contractor will repair any deficiencies that do not meet the performance guideline.

2–6–4   **Observation:** A steel post/column is out of plumb.

Performance Guideline: Steel posts/column should not be out of plumb in excess of ⅜ inch in 8 feet when measured vertically.

Corrective Measure: The contractor will repair any deficiencies that do not meet the performance guideline.

Interior Floor Construction

Floor System

3–1–1   **Observation:** Springiness, bounce, shaking, or visible sag is present in the floor system.

Performance Guideline: All *beams, joists, headers,* and other dimensional or manufactured structural members will be sized according to the manufacturers' specifications or prevailing building codes.

Corrective Measure: The contractor will reinforce or modify, as necessary, any member of the floor system not meeting the performance guideline.

Discussion: *Deflection* may indicate insufficient stiffness in the lumber or may reflect an aesthetic consideration independent of the strength and safety requirements of the lumber. Structural members are required to meet standards for both stiffness and strength. If a consumer expresses a preference to the contractor before construction, the contractor and the consumer may agree upon a higher standard of deflection.

Beams, Columns, and Posts

3–2–1   **Observation:** An exposed wood column, post, or beam is split.

Performance Guideline: Sawn wood columns, posts, or beams will meet the grading standard for the species used at the span and load as prescribed in the applicable building code.

Corrective Measure: The contractor will repair or replace any wood column, post, or beam that does not meet the performance guideline. Filling splits with appropriate filler is an acceptable method of repair.

Discussion: Columns, posts, and beams will sometimes split as they dry after installation. Splitting is acceptable and is not a structural concern if the columns, posts, or beams have been sized according to manufacturer's specifications or applicable building codes. Some materials have inherent cracks or imperfections; these do not require repair.

3–2–2   **Observation:** An exposed wood beam or post is twisted or bowed.

Performance Guideline: Exposed wood posts and beams will meet the grading standard for the species used. Posts and beams with bows and twists exceeding ¾ inch in an 8-foot section are considered excessive.

Corrective Measure: The contractor will repair or replace any beam or post with a bow or twist that exceeds the performance guideline.

Discussion: Beams and posts, especially those 3½ inches or greater in thickness (which normally are not kiln dried) will sometimes twist or bow as they dry after milling or installation. Twisting or bowing is usually not a structural concern if posts and beams have been sized according to manufacturers' specifications or applicable building codes.

3–2–3   **Observation:** An exposed wood beam or post is cupped.

Performance Guideline: Cups exceeding ¼ inch in 5½ inches are considered excessive.

Corrective Measure: The contractor will repair or replace any beam or post with a defect that does not meet the performance guideline.

Discussion: Cupped lumber is lumber that has warped or cupped across the grain in a concave or convex shape. Beams and posts, especially those 3½ inches or greater in thickness (which normally are not kiln dried), may cup as they dry after milling or installation and is not a defect.

Subfloor and Joists

3–3–1   **Observation:** The wood *subfloor* squeaks or seems loose.

Performance Guideline: Although a totally squeak-proof floor cannot be guaranteed, frequent, loud squeaks caused by improper installation or loose subflooring are deficiencies.

Corrective Measure: The contractor will refasten or take other corrective action of any improperly installed or loose subfloor to attempt to reduce squeaking to the extent possible within reasonable repair capability without removing floor or ceiling finishes.

Discussion: There are many possible causes of floor squeaks. One of the more common sources of squeaks is wood moving along the shank of a nail. Squeaking frequently occurs when lumber, floor sheathing, or boards move slightly when someone walks over them. Boards and floor sheathing may become loose due to shrinkage of the floor structure or subfloor as it dries after installation or seasonal changes in temperature and humidity. Nails used to fasten metal connectors (*joist hangers,* tie-down straps, etc.) may cause squeaks. The nature of wood and construction methods makes it practically impossible to eliminate all squeaks during all seasons. Fastening loose subflooring with casing nails into carpet and counter sinking the head is an acceptable method of repair. Snap-off screws may also be used to refasten subflooring through carpet. If there is no finish ceiling below, repairs can be made by shimming or other techniques to repair squeaks are an acceptable alternative.

3–3–2   **Observation:** A wood subfloor is uneven.

Performance Guideline: Subfloors should not have more than a ¼ inch ridge or depression within any 32-inch measurement. Measurements should not be made at imperfections that are characteristic of the material used. This guideline does not cover transition points between different materials.

> **Remodeling Specific Guideline:** The consumer and the contractor may agree to build a wood floor not level to match or otherwise compensate for preexisting conditions.

Corrective Measure: The contractor will correct or repair the subfloor to meet the performance guideline.

3–3–3   **Observation:** A wood subfloor is not level.

Performance Guideline: The floor should not slope more than ½ inch in 20 feet. Crowns and other lumber characteristics that meet the standards of the applicable grading organization for the grade and species used are not defects. Deflections due to overloading by the consumer are not the contractor's responsibility.

> **Remodeling Specific Guideline:** The contractor and the consumer may agree to build an addition not level to keep the floor of an addition on the same plane, and/or the roof ridge on the same line, as those of an existing, out-of-level structure, or to compensate for some other preexisting condition.

Corrective Measure: The contractor will make a reasonable and cost-effective effort to modify the floor to comply with the performance guideline.

Discussion: Sloped floors have both an aesthetic and functional consideration. Measurements for slope should be made across the room, not in a small area.

3–3–4   **Observation:** Deflection and/or flex is observed in a floor system constructed of wood I-joists, floor trusses, or similar products.

Performance Guideline: All wood I-joists and other manufactured structural components in the floor system will be sized and installed as provided in the manufacturers' instructions and applicable building codes.

Corrective Measure: The contractor will reinforce or modify as necessary any floor component that does not meet the performance guideline.

Discussion: Some deflection and/or flex is normal and is not an indication of deficiency in the strength and safety of the product. If a consumer requests it, the contractor and consumer may agree to more stringent criteria in writing prior to construction. Code requirements allow for deflection under load that includes human traffic and at "rest" which may seem more springy than solid sawn lumber.

3–3–5   **Remodeling Specific Observation:** Wood flooring is not level at the transition of an existing floor to a room addition floor.

Performance Guideline: Flooring at a transition area should not slope more than ⅛ inch over 6 inches unless a threshold is added. Overall step-down, unless previously agreed upon with the consumer, should not exceed 1⅛ inches. Variations caused by seasonal or temperature changes are not a defect.

Corrective Measure: The contractor will correct the floor transition to meet the performance guideline.

Discussion: All wood members shrink and expand seasonally, with variations in temperature and humidity, and with aging. After installation, dimensional lumber can shrink up to ½ inch for some boards. If the flooring, subfloor, or underlayment was not purposely overlapped onto the existing floor, the resulting irregularity is not a defect, but a natural result and characteristic of the wood's aging process. Either the old or the new floors may slope along the floor joist span. Joists in an older home may have deflected under load. This and other conditions may cause a hump at the juncture of the old to new.

3–3–6 **Remodeling Specific Observation:** The floor pitches to one side in the door opening between the existing construction and the addition.

Performance Guideline: If the pitch is the result of the floor of the existing dwelling not being level, in most situations a transition threshold is an appropriate and acceptable means of addressing the preexisting condition.

Corrective Measure: The contractor will use best judgment in making a reasonable and cost-effective effort to meet the performance guideline.

Discussion: If the difference between the existing home and addition was not evident at the time of construction or consumer did not express a transition-based solution, adding a transition threshold may incur additional costs.

3–3–7 **Observation:** Exterior sheathing or subfloor materials have delaminated or swollen.

Performance Guideline: Subfloor and exterior sheathing for the surfaces upon which finish materials will be attached/laid shall be flat enough and strong enough to prevent failure of finish materials on that side of the material. The non-finish side shall be based on strength of material only.

Corrective Measure: Defective materials shall be replaced or repaired.

Discussion: If replacement of finish materials is necessary, it shall be done to match existing finishes as closely as practical. Some swelling may occur during construction due to the added moisture from concrete, joint compound, primers and paints, which may remain until the space is closed in and heat is available. Some swelling can be removed by sanding or planing to return the edging and swollen areas to a more uniform condition. Sheathing with delamination on more than ½ of the plys or ¼ of the OSB thickness should be replaced, otherwise the panels have enough strength to not cause structural issues.

4

Walls & Ceilings

Wall Framing

4–1–1   **Observation:** A wood-framed wall is not plumb.

Performance Guideline: The interior face of wood-framed walls should not be more than ⅜ inch out of plumb for any 32 inches in any vertical measurement.

Remodeling Specific Guideline: The contractor and consumer may agree to intentionally build walls out of plumb to match the existing structure to accommodate or compensate for inaccuracies in the existing structure, and to disregard the performance guideline to match a preexisting structural condition.

Corrective Measure: The contractor will correct the wall to meet the performance guideline.

4–1–2   **Observation:** The wall or ceiling is bowed.

Performance Guideline: Walls and ceilings should not bow within the warranty period by more than ½ inch out of line within any 32-inch horizontal measurement, or ½ inch out of line within any 8-foot vertical measurement. This vertical or horizontal measurement shall be taken a minimum of 16 inches from any drywall or plaster corner or opening.

Remodeling Specific Guideline: If new wall cladding is installed on existing framed walls, the contractor and consumer may agree to straighten the wall as part of the scope of work, or to install new cladding over existing framing, and to disregard the performance guideline to match a preexisting structural condition.

Corrective Measure: The contractor will repair the wall to meet the performance guideline.

Discussion: All interior and exterior walls have slight variances in their finished surface. On occasion, the underlying framing may warp, twist, or bow after installation, which is not a structural deficiency.

4–1–3 **Observation:** Deflection is observed in a beam, header, girder, or other dimensional or manufactured structural member in a wall.

Performance Guideline: All beams, headers, girders, and other dimensional or manufactured structural members in the wall system will be sized according to the manufacturers' specifications and applicable building codes that allow for specified amounts of deflection.

Corrective Measure: The contractor will reinforce or modify, as necessary, any beam, header, girder, or other dimensional or manufactured structural member in the wall system that does not meet the performance guideline.

4–1–4 **Observation:** Warping, checking or splitting of wood framing which materially affects its intended purpose.

Performance Guideline: If a condition exists where checking, splitting or warping materially affects the structural integrity of the individual framing member or any contractor-applied surface material attached thereto, then that condition shall be remedied.

Corrective Measure: Contractor will repair, replace, or stiffen the frame member as needed.

Moisture Barriers and Flashing

4–2–1 **Observation:** *Bulk moisture* movement (or liquid flow) is penetrating around a window or door and is visible from inside the home.

Performance Guideline: Windows and doors should be installed and flashed in accordance with manufacturer's specifications and/or as required by prevailing building codes.

Corrective Measure: The contractor will correct to meet the performance guideline.

Discussion: Windows and doors are not completely water resistant. They always (except fixed windows and doors) have cracks or joints through which, with enough wind pressure, wind-driven rain can penetrate. The wind rating specifications for windows and doors are higher than the water rain events, such as short-term intense thunderstorms and tropical storms, because water can be expected to penetrate windows and doors. The consumer is responsible for keeping *weep holes* clean of debris as they are designed to allow wind driven rain to be diverted from the window sill. Any consumer applied caulking that causes water to not weep out would not require corrective measures by the contractor.

4–2–2   **Observation:** An exterior wall leaks because of improper *caulking* installation or failure of the caulking material.

Performance Guideline: Joints and cracks in exterior wall surfaces and around openings should be protected and/or caulked to prevent the entry of water.

Corrective Measure: <u>One time only</u> during the warranty period, the contractor will repair or caulk joints and cracks with exterior grade caulk, as necessary, to correct deficiencies.

Discussion: Even when properly installed, caulk eventually will shrink and crack. Maintenance of caulk is the consumer's responsibility.

Insulation

4–3–1   **Observation:** Insulation is insufficient.

Performance Guideline: The contractor should install insulation according to R-values designated in the contract documents or as required by the prevailing building code.

Corrective Measure: The contractor will install insulation to meet the performance guideline.

4–3–2   **Observation:** Foam or cellulose insulation appears to sag or shrink away from the cavity during the warranty period.

Performance Guideline: Shrinkage/sagging should not be more than ½ inch at the top and ⅛ inch on sides.

Corrective Measure: The contractor will correct insulation to meet the performance guideline.

Discussion: Some space is created by the shrinkage of the framing members and not the insulation and is both expected and acceptable.

4–3–3   **Observation:** Insulation around interior penetrations is either lacking or allows noticeable air flow.

Performance Guideline: Penetrations shall be treated with insulation, air barrier membrane or moisture barrier materials to prevent conditioned air to pass.

Corrective Measure: The contractor will correct to prevent air movement around penetrations.

Windows and Glass

4–4–1   **Observation:** A window is difficult to open or close.

Performance Guideline: Windows should require no greater operating force than that described in the manufacturer's specifications.

Remodeling Specific Guideline: Windows covered by the contractor that are inoperable or operable at greater than stated force are the contractor's responsibility.

Corrective Measure: The contractor will correct or repair the window as required to meet the performance guideline.

4–4–2   **Observation:** Window glass is broken and/or a screen or window hardware is missing or damaged.

Performance Guideline: Glass should not be broken and screens and hardware should not be damaged or missing at the time of substantial completion of the project. Only screens included in the original contract will be installed.

Corrective Measure: Broken glass, missing or damaged screens, or missing or damaged hardware reported to the contractor prior to substantial completion of the project will be installed or replaced. Broken glass, missing or damaged screens, or missing or damaged hardware reported after substantial completion of the project are the consumer's responsibility.

4–4–3   **Observation:** Water is observed in the home around a window unit during or after rain.

Performance Guideline: Window installation should be performed in accordance with manufacturer's specifications so that water does not intrude beyond the drainage plane of the window during normal rain conditions. Windows should resist water intrusion as specified by the window manufacturer.

Corrective Measure: The contractor will correct any deficiencies attributed to improper installation. Any deficiencies attributed to the window unit's performance will be addressed by the window manufacturer's warranty.

Discussion: Leakage at the glazing interface is covered under the manufacturer's warranty. Windows have a limited ability to resist excessive wind-driven rain but should perform according to manufacturer's specifications.

The consumer is responsible for keeping weep holes clean of debris as they are designed to allow wind-driven rain to be diverted from the windowsill.

4–4–4 **Observation:** Window *grids, grilles,* or *muntins* fall out or become out of level.

Performance Guideline: Window grids, grilles, or muntins should not disconnect, fall, or become out of level.

Corrective Measure: <u>One time only</u> during the warranty period, window grids, grilles, or muntins will be repositioned, repaired, or replaced.

4–4–5 **Observation:** Glass surfaces are scratched.

Performance Guideline: Glass surfaces should not have scratches visible from 10 feet under *normal lighting* conditions at the time of substantial completion of the project.

> **Remodeling Specific Guideline:** This guideline does not apply to existing windows unless they are part of the remodeling contract or are damaged by the contractor. The contractor and consumer should examine existing windows prior to contract execution.

Corrective Measure: The contractor will repair or replace any scratched glass surface if noted prior to substantial completion of the project.

4–4–6 **Observation:** Double hung windows do not stay open.

Performance Guideline: Windows should stay within a 2 inch tolerance up or down when placed in an open position.

Corrective Measure: <u>One time only</u> during the warranty period, Contractor shall adjust balances and show consumer that method of adjustment for future client use.

Discussion: A too tightly sealed window may become too difficult to lift or one not sealed enough may fall or rise on its own.

4–4–7 **Observation:** Condensation or frost appears on window frames or glass panes.

Performance Guideline: Windows and doors installed in accordance with the manufacturer's instructions and the prevailing building codes may exhibit condensation or frost.

Corrective Measure: No corrective action is required by the contractor.

Discussion: Condensation usually results from conditions beyond the contractor's control. Moisture in the air can condense into water and collect on cold surfaces, particularly in the winter months when the outside temperature is low. Blinds and drapes can prevent air within the home from moving across the cold surface and picking up the moisture. Occasional condensation on windows and doors in the kitchen, bath, or laundry area is also common. It is the consumer's responsibility to maintain proper humidity by properly operating heating and cooling systems' exhaust fans and allowing moving air within the home to flow over the interior surface of the windows. In hot, humid climates, condensation can occur on the outside of windows when the outdoor humidity is especially high (in early mornings when windows are cool). Air conditioning vents are usually aimed at windows and glass doors to maximize comfort and can cause surface condensation.

Exterior Doors

4–5–1   **Observation:** An exterior door is warped.

Performance Guideline: Exterior doors should not warp to the extent that they become inoperable or cease to be weather-resistant. A ¼-inch tolerance as measured diagonally from corner to corner is acceptable.

Corrective Measure: The contractor will correct or replace exterior doors that do not meet the performance guideline.

Discussion: Most exterior doors will warp to some degree due to the difference in the temperature and humidity between inside and outside surfaces; ¼ inch across the plane of the door measured diagonally from corner to corner is an acceptable tolerance. Warping may also be caused by improper or incomplete finishing of the door including sides, top, and bottom. The contractor is not responsible for warpage if painting of doors is not within the contractor's scope of work.

4–5–2   **Observation:** Raw wood shows at the edges of an inset panel inserted into a wood exterior door during the manufacturing process.

Performance Guideline: This is a common occurrence in wood doors with panels.

Corrective Measure: Since this occurrence is common, no corrective action is required by the contractor.

Discussion: Wood products expand and contract with changes in temperature and humidity. Wooden inserts are intentionally loosely fitted into the rails by the manufacturer to allow the inserts to move, which minimizes splitting of the panel or other damage to the door.

4–5–3   **Observation:** A wooden door panel is split.

Performance Guideline: A split in a panel should not allow light to be visible through the door.

Corrective Measure: <u>One time only</u> during the warranty period, the contractor will repair and paint or stain the split panel that does not meet the performance guideline. Caulking and fillers are acceptable.

Discussion: Wooden inserts are loosely fitted into the door to allow the inserts to move, which minimizes splitting of the panel or other damage to the door. On occasion, a panel may become "locked" by paint or expansion of the edges with changes in temperature and humidity and no longer "float" between the rails. This may result in the panel splitting. The repainted area may not blend with the remainder of the door or other doors on the home.

4–5–4   **Observation:** An exterior door sticks or binds.

Performance Guideline: Exterior doors should operate smoothly, except a door may stick during periods of high humidity or with variations in temperature.

Corrective Measure: The contractor will adjust or replace the door to meet the performance guideline if the problem is caused by faulty workmanship or materials.

Discussion: Exterior doors may warp or bind to some degree because of the difference in the temperature and/or humidity between inside and outside surfaces. The contractor is not responsible for warpage if painting of doors was not within the contractor's scope of work. Any changes to originally installed door hardware, *weather stripping* or other door components that cause improper operation are not the contractor's responsibility.

4–5–5   **Observation:** An exterior door will not close and latch.

Performance Guideline: Exterior doors should close and latch.

Corrective Measure: <u>One time only</u> during the warranty period, the contractor will adjust the door or latching mechanism to meet the performance guideline.

Discussion: Exterior doors may warp or bind to some degree because of the difference in the temperature, humidity, or both, between inside and outside surfaces. Latching also can be affected by natural settling. Subsequent adjustments may be necessary by the consumer. The contractor is not responsible for warpage if painting of doors was not within the contractor's scope of work.

4–5–6   **Observation:** The plastic molding on the primary door behind the storm door droops/melts from exposure to sunlight.

Performance Guideline: It is a common occurrence for the plastic molding behind storm doors to droop/melt.

Corrective Measure: No corrective action is required by the contractor.

Discussion: Plastic moldings may melt or deform if the exterior door is covered by a storm door during a warm season, or if it faces the sun. This is not a defect of the door, but a problem caused by the trapping of heat between the primary door and the storm door. The storm door, or if a combo unit, the storm panel should be removed and reinstalled by the consumer as a part of normal seasonal maintenance (i.e., removed in the spring and reinstalled in the fall). The consumer is also cautioned to follow the manufacturer's recommendations on painting the moldings. Dark colors will tend to absorb more heat.

4–5–7   **Observation:** Caulking or glazing on the primary door behind the storm door cracks or peels.

Performance Guideline: It is a common occurrence for caulking or glazing on the primary door behind the storm door to crack or peel.

Corrective Measure: No corrective action is required by the contractor.

Discussion: High temperatures may cause glazing and caulking to harden and/or fail prematurely if the door is covered by a storm door during a warm season or if it faces the sun. This is not a defect of the door, caulking, or glazing, but a problem caused by the trapping of heat between the primary door and the storm door. The storm door, or if a combo unit, the storm panel should be removed and reinstalled by the consumer as a part of normal seasonal maintenance (i.e., removed in the spring and reinstalled in the fall). The consumer is also cautioned to follow the manufacturer's recommendations on painting the moldings. Dark colors will tend to absorb more heat.

4–5–8   **Observation:** A door swings open or closed by the force of gravity.

Performance Guideline: Exterior doors should not swing open or closed by the force of gravity alone.

> **Remodeling Specific Guideline:** This guideline does not apply where a new door is installed in an existing wall that is out of plumb.

Corrective Measure: The contractor will adjust the door to prevent it from swinging open or closed by the force of gravity.

4–5–9   **Observation:** The reveal around an exterior door edge, doorjamb, and/or threshold is uneven.

Performance Guideline: Gaps between adjacent components should not vary by more than $3/16$ inch along each side of the door.

> **Remodeling Specific Guideline:** This guideline does not apply where a new door is installed in an existing wall that is out of plumb or an existing opening that is out of square.

Corrective Measure: The contractor will repair the existing unit to meet the performance guideline.

Discussion: Doors must have gaps at their perimeter to accommodate expansion/contraction due to variations in temperature and/or humidity and to enable the door to operate over a wide range of environmental conditions.

4–5–10   **Observation:** Air movement or light is observed around a closed exterior door.

Performance Guideline: Weather stripping will be installed and sized properly to seal the exterior door when closed in order to prevent excessive air infiltration.

Corrective Measure: The contractor will adjust exterior door unit or weather stripping to meet the performance guideline.

Discussion: Doors must have gaps at their perimeter to accommodate expansion/contraction due to variations in temperature and/or humidity and to enable the door to operate over a wide range of environmental conditions. Weather stripping seals the gaps required for proper operations to prevent excessive air infiltration. At times of high wind or temperature differentials inside the home and outside, there may be noticeable air movement around a closed door's perimeter. A small glimmer of light seen at the corners of the door unit is normal. Weather stripping should be kept cleaned and maintained by the consumer.

4–5–11   **Observation:** Exterior door hardware or kickplate has tarnished.

       Performance Guideline: Finishes on door hardware or kickplates installed by the contractor are covered by the manufacturer's warranty.

       Corrective Measure: No corrective action is required by the contractor.

4–5–12   **Observation:** A sliding patio door or screen does not stay on track.

       Performance Guideline: Sliding patio doors and screens should slide properly on their tracks at the time of substantial completion of the project. The cleaning and maintenance necessary to preserve proper operation are consumer responsibilities.

       Corrective Measure: One time only during the warranty period, the contractor will adjust the door or screen.

       Discussion: Proper operation should be verified by the consumer and the contractor at the time of substantial completion of the project.

4–5–13   **Observation:** A sliding patio door does not roll smoothly.

       Performance Guideline: Sliding patio doors should roll smoothly at the time of substantial completion of the project. The cleaning and maintenance necessary to preserve proper operation are consumer responsibilities.

       Corrective Measure: One time only during the warranty period, the contractor will adjust the door.

       Discussion: Proper operation should be verified by the consumer and the contractor at the time of substantial completion of the project.

4–5–14   **Observation:** A doorknob, deadbolt, or lockset does not operate smoothly.

Performance Guideline: A doorknob, deadbolt, or lockset should not stick or bind during operation.

Corrective Measure: <u>One time only</u> during the warranty period, the contractor will adjust, repair, or replace knobs that are not damaged by the consumer.

Discussion: Locksets may feel heavy or stiff but are operating as intended by the manufacturer. This can be true for locksets of all price ranges. Electronic locks can be more sensitive than traditional lock sets and may need adjustment. If installed by the contractor, <u>one time only</u> during the warranty period, the contractor will adjust so the lock latches.

4–5–15   **Observation:** Storm doors, windows or screens do not operate or fit properly.

Performance Guideline: Storm doors, storm windows or screens shall work as intended and fit to provide the protection intended.

Corrective Measure: Contractor will adjust, repair, or replace to ensure proper fit and operation.

4–5–16   **Observation:** Condensation or frost appears on exterior door, door glass or frame.

Performance Guideline: Exterior doors installed in accordance with the manufacturer's instructions and the prevailing building codes may exhibit condensation or frost.

Corrective Measure: No corrective action is required by the contractor.

Discussion: Condensation usually results from conditions beyond the contractor's control. Moisture in the air can condense into water and collect on cold surfaces, particularly in the winter months when the outside temperature is low. Blinds and drapes can prevent air within the home from moving across the cold surface and picking up the moisture. Occasional condensation on windows and doors in the kitchen, bath, or laundry area is also common. It is the consumer's responsibility to maintain proper humidity by properly operating heating and cooling systems' exhaust fans and allowing moving air within the home to flow over the interior surface of the windows. In hot, humid climates, condensation can occur on the outside of windows when the outdoor humidity is especially high (in early mornings when windows are cool). Air conditioning vents are usually aimed at windows and glass doors to maximize comfort and can cause surface condensation.

Exterior Finish

Wood and Wood Composite Siding

5–1–1 **Observation:** Siding is bowed.

Performance Guideline: Bows exceeding ½ inch in 32 inches are considered excessive.

Remodeling Specific Guideline: If new wall covering is installed on existing framed walls, the contractor and consumer may agree to straighten out the walls as part of the scope of work. Alternatively, the parties may agree to install new wall covering over existing framing and disregard the performance guideline to match a preexisting structural condition.

Corrective Measure: The contractor will replace any bowed wood siding that does not meet the performance guideline and will finish the replacement siding to match the existing siding as closely as possible.

Discussion: If the siding is fastened by nails driven into studs, expansion caused by changing relative temperatures and/or humidity may cause bulges or waves. Even with proper installation, siding will tend to bow inward and outward in adjacent stud spaces.

5–1–2 **Observation:** An edge or gap is visible between adjacent pieces of siding or siding panels and other materials.

Performance Guideline: Gaps wider than 3⁄16 inch are considered excessive, unless the siding is installed as prescribed by the manufacturer's instructions, which may include the options to caulk as with cement board siding and spacing required for expansion and contraction of composite siding.

Corrective Measure: The contractor will repair gaps that do not meet the performance guideline.

Discussion: Proper repair can be completed by providing joint covers or by caulking the gap. This is important if the gaps were intentionally left at joints for expansion and contraction. If the siding is painted, the contractor will paint the new caulking to match the existing siding as closely as possible, but an exact match cannot be guaranteed or promised.

5–1–3   **Observation:** Siding is not parallel with the course above or below.

Performance Guideline: A piece of siding should not be more than ½ inch off parallel with contiguous courses in any 20-foot measurement.

> **Remodeling Specific Guideline:** The contractor and consumer may agree to install siding to match existing conditions on the existing structure and to disregard the performance guideline for this item. If the contractor and consumer have agreed that the floor of an addition is to be on a different plane from an existing floor (i.e., out of level), the siding on the addition may not be parallel and in line with the existing siding.

Corrective Measure: The contractor will reinstall siding to meet the performance guideline for straightness and will replace with new siding any siding damaged during removal.

5–1–4   **Observation:** Face nails have been driven below the surface of wood composite siding.

Performance Guideline: Siding nails should be driven in accordance with the manufacturer's installation instructions.

Corrective Measure: The contractor will repair as necessary to meet the performance guideline by filling with appropriate filler. Touch-up paint may not match the surrounding area.

5–1–5   **Observation:** Siding boards have buckled, warped, or cupped.

Performance Guideline: Boards that project more than 3/16 inch within 5½ inches of siding are considered excessive.

Corrective Measure: The contractor will repair or replace any boards that do not meet the performance guideline.

Discussion: Buckling, warping, or cupping is caused by wood expanding because of increased temperature, relative humidity, or both.

5–1–6   **Observation:** Siding boards have split.

Performance Guideline: Splits wider than ⅛ inch and longer than 1 inch are considered excessive.

Corrective Measure: The contractor will repair siding boards that do not meet the performance guideline by filling with appropriate filler. Touch-up paint may not match the surrounding area.

5–1–7   **Observation:** Wood siding, shakes, or shingles have bled through paint or stain applied by the contractor.

Performance Guideline: Resins and extractives bleeding through paint or stain, or blackening of siding, shakes, or shingles is considered normal, and is especially noticeable if natural weathering, white paint, or semi-transparent stain is specified for the project.

Corrective Measure: No corrective action is required by the contractor.

5–1–8   **Observation:** Siding has delaminated.

Performance Guideline: Siding should not delaminate.

Corrective Measure: Delaminating of siding is covered under the manu-facturer's warranty unless the delaminating was caused by the consum-er's actions or negligence. After substantial completion of the project, the consumer should contact the manufacturer for warranty coverage. Delaminated siding installed by the contractor shall be fixed at the time of substantial completion of the project.

5–1–9   **Observation:** Nail stains are visible on siding or ceiling boards.

Performance Guideline: Stains exceeding ½ inch from the nail which are readily visible from a distance of more than 20 feet are considered excessive.

Corrective Measure: The contractor will remove stains that do not meet the performance guideline.

Discussion: Stains can be caused by oxidation of nails or leaching of extractives from the wood. Use of galvanized nails (even double hot dipped) may not necessarily prevent staining.

Aluminum or Vinyl Siding

5–2–1   **Observation:** Aluminum or vinyl siding is bowed or wavy.

Performance Guideline: Some waviness in aluminum or vinyl siding is expected. Waves or similar distortions in aluminum or vinyl lap siding are considered excessive only if they exceed ½ inch within 32 inches of siding.

If the new siding is applied to existing walls as part of a remodel, the siding will follow the underlying surface. Prior to commencement of the work, the contractor and consumer should agree on whether the existing surface is to be modified.

Corrective Measure: The contractor will correct any waves or distortions to comply with the performance guideline by reinstalling or replacing siding, as necessary. If the walls will not be modified, the Performance guideline will not apply, and the remodeler will take no corrective measure.

5–2–2   **Observation:** Siding is faded.

Performance Guideline: Any color siding, when exposed to the ultraviolet rays of the sun, will fade. Fading cannot be prevented by the contractor. However, panels installed on the same wall and under the same conditions should fade at the same rate.

Corrective Measure: No corrective action is required by the contractor. The consumer should contact the siding manufacturer for issues with inconsistent fading.

Discussion: Color warranties are provided by the siding manufacturer. The consumer should contact the manufacturer with questions or claims regarding changes in color of vinyl or aluminum siding. Color and fade imperfections beyond an expected degree may be covered by the manufacturer's warranty, except where siding is shaded differently from the rest of the wall, such as under shutters or behind vegetation.

5–2–3   **Observation:** Aluminum or vinyl siding trim is loose.

Performance Guideline: Trim should not separate from the home by more than ¼ inch.

Corrective Measure: The contractor will reinstall trim as necessary to comply with the performance guideline.

Discussion: Vinyl siding and accessories should not be caulked in most circumstances, as caulking could impact the product's contraction and expansion characteristics.

5–2–4   **Observation:** Aluminum or vinyl siding courses are not parallel with eaves or wall openings.

Performance Guideline: Any piece of aluminum or vinyl siding more than ½ inch off parallel in 20 feet with a break such as an eave or wall opening is considered excessive.

Remodeling Specific Guideline: If the contractor and consumer agree that the floor of an addition is to be on a different plane from the existing floor (e.g., a preexisting out-of-level condition), the siding on the addition may not be parallel and in line with existing siding. The contractor and consumer may agree to install siding to match existing conditions on the existing structure and to disregard the performance guideline for this item.

Corrective Measure: The contractor will reinstall siding to comply with the performance guideline and will replace with new siding any siding damaged during removal.

5-2-5   **Observation:** Nail heads show in aluminum or vinyl siding.

Performance Guideline: No nail heads in the field of the siding should be exposed.

Corrective Measure: The contractor will install trim as necessary to cover the nails and will install proper trim accessories to avoid face nailing.

Discussion: Vinyl siding generally should not be face nailed. However, there are appropriate and typical occasions when a single face nail may be needed to reinforce a joint or fasten the siding to the wall when it is cut to fit around window frames, doors, roofs, or other obstructions on the wall.

In most cases (the only exception would be the top piece on a gable end), vinyl siding should not be face nailed when proper accessory products are used. For example, under a window application the contractor can use the J-channel trim and utility trim, and snap punch the top of the vinyl siding. If face nailing is the only option, the contractor should predrill a ⅛ inch diameter hole to allow for expansion and contraction.

5-2-6   **Observation:** Aluminum or vinyl siding trim accessory is loose from caulking at windows or other wall openings.

Performance Guideline: Siding trim accessories should not separate from caulking at windows or other wall openings during the warranty period.

Corrective Measure: One time only during the warranty period, the contractor will repair or caulk, as necessary, to eliminate the separation.

5-2-7   **Observation:** Aluminum or vinyl siding has gaps or inconsistent cuts.

Performance Guideline: Gaps will comply with the manufacturer's guidelines, but cuts should be concealed by trim. Field cut edges of vinyl or aluminum siding should not be visible when proper trim and accessories are used.

39

Remodeling Specific Guideline: The contractor and consumer may agree to install siding to match conditions on the existing structure and to disregard the performance guideline for this item.

Corrective Measure: The contractor will ensure that the appropriate trim/accessory is installed to eliminate potentially revealing site cuts. If cuts in siding panels are so uneven that they are not concealed by trim, the panel will be replaced.

5–2–8   **Observation:** Aluminum or vinyl siding is not correctly spaced from moldings.

Performance Guideline: Prescribed spacing between siding and accessory trim is typically ¼ inch, or should comply with the manufacturer's installation instructions.

Remodeling Specific Guideline: The contractor and consumer may agree to install siding to match conditions on the existing structure and to disregard the performance guideline for this item.

Corrective Measure: The contractor will correct the spacing to meet the performance guideline.

5–2–9   **Observation:** Aluminum or vinyl siding is rattling or banging on house.

Performance Guideline: Prescribed nailing for siding should meet manufacturers installation specifications.

Corrective Measure: The contractor will correct the nailing to meet the performance guideline.

Discussion: During times of high winds the siding will have some movement and there will be noise from the siding against the house.

Fiber Cement Board Siding

5–3–1   **Observation:** Fiber cement board siding is cracked or chipped.

Performance Guideline: As a cement product, this siding is susceptible to the same characteristic limitations as other cement products. Cracks more than 2 inches in length and ⅛ inch in width are considered excessive. Chips or dents not reported at the time of substantial completion of the project are not covered.

Corrective Measure: Cracked or chipped cement board will be repaired or replaced as necessary, as determined by the contractor.

Discussion: The manufacturer's instructions include guidelines to reduce chipping or cracking of siding.

5–3–2   **Observation:** Cement board siding is improperly fastened.

Performance Guideline: Siding should be nailed flush and perpendicular per the manufacturer's instructions. Staples should not be used.

Corrective Measure: The contractor will correct or repair improperly fastened boards. Overdriven nail heads or nails driven at an angle can be filled with siding manufacturer's specified product.

5–3–3   **Observation:** Cement board siding and trim have visible gaps.

Performance Guidelines: Fiber cement siding shall have a ⅛ inch gap between trims of windows and doors. Install all butt ends and joints in contact with one another.

Corrective Measure: Contractor will repair siding to meet the performance guidelines.

Discussion: All fiber cement installation, caulking should be in accordance with each manufacturer's instructions and specifications.

Masonry and Veneer

5–4–1   **Observation:** A masonry or veneer wall or mortar joint is cracked.

Performance Guideline: Cracks visible from distances in excess of 20 feet, or more than ¼ inch in width are not acceptable.

Corrective Measure: The contractor will repair cracks in excess of the performance guideline by tuck pointing, patching, or painting, as deemed most appropriate by the contractor.

Discussion: Hairline cracks resulting from shrinkage and cracks due to minor settlement are common in masonry or veneer walls and mortar joints, and do not necessarily represent a defect. An exact match of mortar after a repair cannot be guaranteed.

5–4–2 **Observation:** Cut bricks below openings in masonry walls are of different thickness.

Performance Guideline: Cut bricks used in the course directly below an opening should not vary from one another in thickness by more than ¼ inch. The smallest dimension of a cut brick should be greater than 1 inch.

Corrective Measure: The contractor will repair the wall to meet the performance guideline.

Discussion: Bricks are cut to achieve required dimensions at openings and ends of walls when it is not possible to match unit/mortar coursing. An exact match of brick and mortar after a repair cannot be guaranteed.

5–4–3 **Observation:** A brick course is not straight.

Performance Guideline: No point along the bottom of any course will be more than ¼ inch higher or lower than any other point within 10 feet along the bottom of the same course, or ½ inch in any course length.

> **Remodeling Specific Guideline:** The contractor and consumer may agree to install brick veneer to match conditions on the existing structure and to disregard the performance guideline for this item.

Corrective Measure: The contractor will rebuild the wall as necessary to meet the performance guideline.

Discussion: Dimensional variations of the courses depend upon the variations in the brick selected. An exact match of brick and mortar after a repair cannot be guaranteed.

5–4–4 **Observation:** Brick veneer is spalling.

Performance Guideline: Spalling of newly manufactured brick should not occur and is considered excessive. Spalling of used manufactured brick is acceptable.

Corrective Measure: Spalling of newly manufactured brick is covered by the manufacturer's warranty. No corrective action is required by the contractor.

5–4–5 **Observation:** Mortar stains are observed on exterior brick or stone.

Performance Guideline: Exterior brick and stone should be free from mortar stains detracting from the appearance of the finished wall when viewed from a distance of 20 feet.

Corrective Measure: The contractor will clean the mortar stains to meet the performance guideline.

5–4–6 **Observation:** Efflorescence is present on the surface of masonry or mortar.

Performance Guideline: This is a common condition caused by moisture reacting with the soluble salts in the mortar.

Corrective Measure: No corrective action is required by the contractor.

Discussion: Efflorescence is evidenced by the presence of a white film on the surface of masonry or mortar. It is a particularly common occurrence where masonry or concrete are in contact with high moisture levels because masonry products absorb and retain moisture.

5–4–7 **Observation:** There is water damage to interior walls as a result of a leak in the exterior brick or stone during normal weather conditions.

Performance Guideline: Exterior brick and stone walls should be constructed and flashed according to the prevailing building code to prevent water penetration to the interior of the structure under normal weather conditions.

Corrective Measure: The contractor will repair the wall to meet the performance guideline, unless the water damage resulted from factors beyond the contractor's control.

Discussion: Water penetration resulting from external factors such as extreme weather conditions, grading alterations or any landscape alterations by others that raises the grade or impacts the proper drainage away from the walls of the structure, or improper use of sprinkler systems are not the contractor's responsibility.

Stucco and Parged Coatings

5–5–1 **Observation:** An exterior stucco wall surface is cracked.

Performance Guideline: Cracks in exterior stucco wall surfaces should not exceed ⅛ inch in width.

Corrective Measure: <u>One time only</u> during the warranty period, the contractor will repair cracks exceeding ⅛ inch in width using an exterior grade caulking or sealant. Caulking and touch-up painting are acceptable. An exact color or texture match may not be attainable.

Discussion: Hairline cracks in stucco or cement plaster (parging) are common, especially if the coatings have been applied directly to masonry back up.

5–5–2 **Observation:** The color, texture, or both, of exterior stucco walls are not uniform.

Performance Guideline: Exterior stucco walls may not match when applied on different days or under differing environmental conditions (e.g., temperature, humidity, etc.).

> **Remodeling Specific Guideline:** The color, texture, or both, of new exterior stucco walls may not match those of old exterior stucco walls.

Corrective Measure: Stucco finishes are unique and an exact match of color, texture, or both may not be practical; therefore, no corrective action is required by the contractor.

Discussion: Stucco includes cement-based coatings and similar synthetically based finishes. Several variables affect coloring and texture of stucco. It is difficult, if not impossible, to achieve a color match between stucco coatings applied at different times. Approved samples prior to installation can minimize misunderstandings about color and texture.

5–5–3 **Observation:** Coating has separated from the base on an exterior stucco wall.

Performance Guideline: The coating should not separate from the base on an exterior stucco wall.

Corrective Measure: The contractor will repair areas where the coating has separated from the base in accordance with the performance guideline, unless the damage resulted from factors beyond the contractor's control.

Discussion: A number of variables affect coloring and texture of stucco. It is not possible to achieve an exact color and/or texture match between stucco coatings applied at different times.

5–5–4 **Observation:** Lath is visible through stucco.

Performance Guideline: Lath should not be visible through stucco, nor should the lath protrude through any portion of the stucco surface.

Corrective Measure: The contractor will make necessary corrections to meet the performance building guidelines. The finish color and/or texture may not match.

Discussion: A number of variables affect coloring and texture of stucco. It is not possible to achieve an exact color and/or texture match between stucco coatings applied at different times.

5–5–5   **Observation:** Rust marks are observed on the stucco finish coat.

Performance Guideline: Rust marks on the stucco surface are considered excessive if more than five marks measuring more than 1 inch long occur per 100 square feet.

Corrective Measure: The contractor will repair, replace, or seal the rusted areas of wall.

Discussion: Rusting may be present in more humid climates due to the natural state of sand used in cement-based products which could include metallic components.

5–5–6   **Observation:** The stucco finish coat does not go all the way to the ground.

Performance Guideline: Weep screed or gap between ground and finish coat of stucco is a proper installation requirement.

Corrective Measure: No corrective measure is necessary.

5–5–7   **Observation:** There is water damage to the exterior wall cavity as a result of a leak in the stucco wall system.

Performance Guideline: Stucco walls should be constructed and flashed to prevent water penetration to the interior of the structure under normal weather and water conditions. Damage to the stucco system caused by external factors out of the contractor's control that result in water penetration is not the contractor's responsibility.

Corrective Measure: If water penetration is the result of a system failure and does not result from external factors, the contractor will make necessary repairs to prevent water penetration through the stucco wall system.

Discussion: The contractor is not responsible for water penetration resulting from external factors such as windblown moisture or sprinkler systems.

5–5–8   **Observation:** Efflorescence is present on the surface of stucco finished surfaces.

Performance Guideline: This is a common condition caused by moisture reacting with the soluble salts in the mortar.

Corrective Measure: No corrective action is required by the contractor.

Discussion: Efflorescence is evidenced by the presence of a white film on the surface of masonry or mortar. It is a particularly common occurrence where masonry or concrete are in contact with high moisture levels because masonry products absorb and retain moisture.

Exterior Trim

5–6–1   **Observation:** Gaps show in exterior trim.

Performance Guideline: Joints between exterior trim elements, including siding and masonry, should not be wider than ¼ inch. In all cases, the exterior trim will perform its function of excluding the elements.

Corrective Measure: The contractor will repair open joints that do not meet the performance guideline. Caulking is an acceptable repair.

Discussion: Reasonable attempts will be made to make repairs using products that match the manufacturer's recommended application instructions.

5–6–2   **Observation:** Exterior trim board is split.

Performance Guideline: Splits wider than ⅛ inch and longer than 1 inch are considered excessive.

Corrective Measure: The contractor will repair splits by filling with durable filler. Touch-up painting may not match the surrounding area.

5–6–3   **Observation:** Exterior trim board is bowed or twisted.

Performance Guideline: Bows and twists exceeding ⅜ inch within 8 feet of trim board are considered excessive.

Corrective Measure: The contractor will repair defects that do not meet the performance guideline by refastening or replacing deformed boards. Touch-up painting may not match the surrounding area.

5–6–4   **Observation:** Exterior trim board is cupped.

Performance Guideline: Cups exceeding 3/16 inch per every 5½ inches of trim board width is considered excessive.

Corrective Measure: The contractor will repair defects that do not meet the performance guideline by refastening or replacing cupped boards. Touch-up painting may not match the surrounding area.

Paint, Stain, and Varnish

5–7–1   **Observation:** Exterior painting, staining, or refinishing is needed because the repair work does not match existing exterior finish.

Performance Guideline: Repairs required under these performance guidelines will be finished to match the immediate surrounding areas as closely as possible when viewed under normal lighting conditions from a distance of 20 feet.

Corrective Measure: The contractor will finish repaired areas as indicated, matching as closely as possible.

Discussion: Touch-up painting, staining, or refinishing may not match the surrounding area exactly in color or sheen because the original coating may have been exposed to sunlight, pollution, weather, and other conditions over a period of time.

5–7–2   **Observation:** Exterior paint or stain has peeled, deteriorated, or flaked.

Performance Guideline: Exterior paints and stains should not peel or flake during the first year.

Corrective Measure: If exterior paint or stain has peeled, developed an alligator pattern, or blistered, the contractor will properly prepare and refinish the affected areas and match their color as closely as practical. Where deterioration of the finish affects more than 50 percent of the piece of trim or wall area, the contractor will refinish the affected component according to manufacturer's product application instructions, or in the absence of such instructions, generally accepted trade practices.

5–7–3   **Observation:** Exterior paint or stain has faded.

Performance Guideline: Fading of exterior paints and stains is common. The degree of fading depends on environmental conditions including ultraviolet (UV) exposure.

Corrective Measure: Because fading is a common occurrence in paint and stains, no corrective action is required by the contractor.

47

5–7–4   **Observation:** There is paint or stain overspray on surfaces not intended for paint or stain.

Performance Guideline: Paint or stain overspray on surfaces not intended for paint or stain that is visible at a distance of 6 feet under normal natural lighting conditions is not acceptable.

Corrective Measure: The contractor will clean the affected surfaces without damaging the surface.

5–7–5   **Observation:** Extensive bleeding of knots, wood or pitch stains show through paint on exterior trim and siding.

Performance Guideline: Knots and other pitch-produced defects should not show through during the warranty period.

Corrective Measure: One time only during the warranty period, the wood will be sealed, stain killed and touch-up painted in affected areas to match as closely as possible.

Roof

Note: Remodeling Specific Guideline: *Where applicable, in the following guidelines the contractor is responsible only for areas of the home where work was performed as specified in the contract, and not for the entire home. The contractor shall not be responsible for repairs and patches made to existing roof.*

Roof Structure

6–1–1   **Observation:** The roof ridge has deflected.

Performance Guideline: Deflections greater than 1 inch within 8 feet of the roof ridge is considered excessive.

Corrective Measure: The contractor will repair the affected ridge that does not meet the performance guideline.

6–1–2   **Observation:** A rafter or ceiling joist bows (up or down).

Performance Guideline: Bows greater than 1 inch within 8 feet of rafter or ceiling joist are excessive.

Corrective Measure: The contractor will repair the affected rafters or joists that bow in excess of the performance guideline.

6–1–3   **Observation:** Roof trusses have deflected.

Performance Guideline: All roof trusses and other manufactured structural roof components in the roof system should be sized according to the manufacturer's specifications or structural engineer's requirements as well as prevailing building codes.

Corrective Measure: The contractor will reinforce or modify as necessary any roof truss or other manufactured structural roof components in the roof system to meet the performance guideline.

Discussion: Deflection is a normal condition that is considered as part of the engineering design of the roof trusses and other manufactured structural roof components. Deflection may be an aesthetic consideration independent of the strength and safety requirements of the product. Deflection and truss movement may be visually evident at the joint of *drywall*. This is common and somewhat unavoidable when there are heavy snow loads or when strong winds have blown on the building and roof causing them to move slightly, but enough to crack brittle drywall joints.

6–1–4 **Observation:** Roof trusses have lifted from the adjoining interior walls.

Performance Guideline: Moisture differences between the *upper chord* and *lower chord* (unheated versus adjacent interior heated spaces) may cause the lower chords to move. Deflection is a normal condition that is considered as part of the engineering design of the roof trusses.

Corrective Measure: No corrective action is required by the contractor.

Discussion: Truss uplift is an aesthetic consideration and is independent of the strength and safety requirements of the truss. This situation will be more prevalent in the winter due to greater variance in moisture and temperature in some regions.

Roof Sheathing

6–2–1 **Observation:** Roof sheathing is wavy or appears bowed.

Performance Guideline: Roof sheathing should not bow more than ½ inch per every 2 feet.

> **Remodeling Specific Guideline:** If new sheathing is installed over existing rafters, the sheathing will follow the bows of the existing rafters. The consumer and contractor should agree on whether the rafters are to be straightened. If they are not to be straightened, the performance guideline for this item will be disregarded.

Corrective Measure: The contractor will straighten bowed roof sheathing as necessary to meet the performance guideline.

Discussion: The contractor may install blocking between the framing members to straighten the sheathing. Under certain viewing conditions and light, minor irregularities in the roof sheathing may be observed. This may be particularly apparent on truss framing with asphalt shingles.

6–2–2 **Observation:** Nails or staples are visible through sheathing or boards (decking) at overhangs.

Performance Guideline: The length of nails and staples used to secure roofing materials is determined by the manufacturer's installation instructions and the prevailing building code.

Corrective Measure: No corrective action is required by the contractor.

Discussion: Nails and staples may protrude through sheathing at overhangs. Their appearance is only an aesthetic concern.

Roof Vents

6–3–1 **Observation:** An attic vent or louver leaks.

Performance Guideline: Attic vents and louvers properly installed should not leak. Infiltration of wind-driven rain and snow are not considered leaks and are beyond the contractor's control.

Corrective Measure: The contractor will repair or replace improperly installed vents as necessary to meet the performance guideline.

6–3–2 **Observation:** Roof vents and attic ventilation seem to create inadequate flow.

Performance Guideline: The total roof vent area should meet the requirements of the prevailing building codes.

  Remodeling Specific Guideline: Existing attic ventilation conditions are not the responsibility of the contractor.

Corrective Measure: The contractor will correct roof ventilation as necessary to meet the performance guideline.

Discussion: Attic ventilation can be provided in variety of ways and proper ventilation may be obtained through ridge vents, soffit vents, gable vents, attic fans, fresh air vents or any combination thereof. Some attics are sealed or finished as conditioned space and do not require outside ventilation. It is the consumer's responsibility to keep the vent locations free from obstructions. In high wind-driven rain events (e.g., over 40 mph), rain that is in the wind will flow with the wind into attics. Louvers in gable end vents can keep large drops of water out of attics when wind speeds are low. However, at even moderate wind speeds the wind will carry water drops with it. It is unavoidable because there is no practical mechanism to separate drops (especially very small ones) from air. Small drops can be like trying to separate drops of fog from air. In short-duration rain events, the accumulation of water may be so small as not to be noticed; however, during tropical depressions and hurricanes, the water can accumulate to damaging quantities. The best preventative measure is to completely block gable end vents during such events. Gable end vents with hinged louvers still can allow wind driven rain to enter an attic in damaging quantities. When rain enters an attic, it can accumulate to such an extent on the top of ceiling drywall that the drywall weakens and the ceiling collapses.

Roof Coverings

Note: *There are many kinds of roofing products. For the purpose of the following performance guidelines, regardless of the actual material used, the term "shingle" will be used to refer to all types of roof coverings.*

6–4–1   **Observation:** The roof or flashing leaks.

Performance Guideline: Roofs and flashing should not leak under normal conditions.

Corrective Measure: The contractor will repair any verified roof or flashing leaks not caused by ice buildup, leaves, debris, abnormal weather conditions, or the consumer's actions or negligence.

Discussion: It is the consumer's responsibility to keep the roof drains, gutters, and downspouts free of ice, leaves, and debris.

6–4–2   **Observation:** Ice builds up on the roof.

Performance Guideline: During prolonged cold spells ice is likely to build up on a roof, especially at the eaves. This condition can occur naturally when snow and ice accumulate.

Corrective Measure: No corrective action is required by the contractor.

Discussion: Prevention of ice buildup on the roof is a consumer maintenance item.

6–4–3   **Observation:** Shingles have blown off.

Performance Guideline: Shingles shall be rated for the wind zone of the project and shall be installed in accordance with the applicable prevailing building code and the instructions provided by the manufacturer on the packaging of the shingles.

Corrective Measure: If shingles were not installed per the manufacturer's installation instructions, the contractor will repair or replace shingles that have blown off.

Discussion: Correctly installed shingles are covered by the manufacturer's warranty. The wind rating of shingles is determined for brand-new shingles using tests of questionable accuracy in predicting actual wind performance especially when time elapses. Generally, shingles lose wind resilience with time as short as a few months. Shingles are not regarded as having sealed to one another until they have reached 135 degrees Fahrenheit for at least 18 hours. In warm, sunny weather, these sealing conditions can occur in just a few days, but until then, shingles are vulnerable to wind. Replacement shingles may not match existing shingles.

6–4–4   **Observation:** Shingles are not horizontally aligned.

Performance Guideline: Shingles should be installed according to the manufacturer's installation instructions.

> **Remodeling Specific Guideline:** The consumer and the contractor may agree prior to installation that the horizontal line of shingles on the roof of an addition need not line up with those of the existing structure if the floors (and hence, the eaves and ridge) are not to be built on the same plane.

Corrective Measure: The contractor will remove shingles that do not meet the performance guideline and will repair or replace them with new shingles that are properly aligned.

Discussion: The bottom edge of dimensional shingles may be irregular; the irregularity is an inherent feature of the design. Replacement shingles may not match existing shingles.

6–4–5   **Remodeling Specific Observation:** New shingles do not match existing shingles.

Remodeling Specific Guideline: The color of new shingles may not exactly match the color of the existing shingles because of weathering and manufacturing variations.

Corrective Measure: The contractor is not responsible for precisely matching the color of existing shingles.

6–4–6 **Observation:** Asphalt shingle edges or corners curl or cup.

Performance Guideline: These conditions are a manufacturer's warranty issue.

Corrective Measure: No corrective action is required by the contractor.

6–4–7 **Observation:** Roofing shingles do not overhang the edges of the roof or they hang too far over the edges of the roof.

Performance Guideline: Shingles should be installed according to the manufacturer's instructions and the prevailing building code.

Corrective Measure: The contractor will reposition or replace shingles as necessary to meet the performance guideline.

Discussion: In high-wind areas, shingles may be purposely installed so they do not extend beyond the edges of the roof. This is to reduce the chance of wind picking the edges up.

6–4–8 **Observation:** Shading or a shadowing pattern is observed on a new shingle roof.

Performance Guideline: Shading or shadowing differences may occur on a new roof.

Corrective Measure: No corrective action is required by the contractor.

6–4–9 **Observation:** Asphalt shingles have developed surface buckling.

Performance Guideline: Asphalt shingle surfaces need not be perfectly flat. However, buckling higher than ¼ inch is considered excessive.

Corrective Measure: The contractor will repair or replace the affected shingles to meet the performance guideline.

Discussion: Reasonable time should be given in cooler weather for shingles to warm and lay flat. Replacement shingles may not match existing shingles.

6–4–10 **Observation:** Sheathing nails have loosened from framing and raised the shingles.

Performance Guideline: Nails should not loosen from roof sheathing enough to raise shingles from surface.

Corrective Measure: The contractor will make corrections as necessary to meet the performance guideline.

Discussion: It is not uncommon for nails to withdraw from the framing because of temperature variations. The contractor can re-drive or remove and replace fasteners that withdraw from the framing. Any resulting holes should be sealed or the shingle should be replaced. Consumer is advised that replacement shingles may not match existing shingles.

6–4–11 **Observation:** Roofing nails or fasteners are exposed at the ridge or hip of a roof.

Performance Guideline: Nails and fasteners should be installed according to manufacturer's instructions.

Corrective Measure: The contractor will seal and/or repair areas to meet the performance guideline.

6–4–12 **Observation:** Areas of a shingle roof are stained.

Performance Guideline: Shingles on a roof may stain.

Corrective Measure: Staining on shingles is unavoidable, no corrective action is required by the contractor.

Discussion: Black stains are indicative of mold that is unavoidable even when stain resistant shingles are installed. Whitish stain can be caused by chemicals in the metals of roof vents or the surrounding metal of pipes leaching onto the roof.

6–4–13 **Observation:** Holes from construction activities are found on the roof surface.

Performance Guideline: Holes from construction activities should be flashed or sealed to prevent leaks.

Corrective Measure: The contractor will repair or replace the affected shingles to meet the performance guideline.

6–4–14 **Remodeling Specific Observation:** Existing roof shingles are telegraphing through new shingles.

Remodeling Specific Guideline: Some telegraphing is common when re-roofing over existing roofing.

Corrective Measure: Because this is an expected and common unavoidable occurrence, no corrective action is required by the contractor.

6–4–15   **Observation:** Water is trapped under membrane roofing.

Performance Guideline: Water should not become trapped under membrane roofing.

Corrective Measure: If water becomes trapped under membrane roofing, the contractor will repair or replace the roofing as necessary to meet the performance guideline.

6–4–16   **Observation:** Membrane roofing is blistered but does not leak.

Performance Guideline: Surface blistering of membrane roofing is caused by conditions of heat and humidity acting on the membrane and is a common occurrence.

Corrective Measure: No corrective action is required by the contractor.

6–4–17   **Observation:** There is standing water on a flat roof.

Performance Guideline: Water should drain from a flat roof, except for minor ponding, within 24 hours of a rainfall or according to manufacturer's specifications.

Corrective Measure: The contractor will take corrective action to meet the performance guideline.

6–4–18   **Observation:** There are visible defects in steep slope roofing.

Performance Guideline: Defects in the roofing surfaces resulting from the contractor's installation that allow water penetration or are easily visible from the ground are the contractor's responsibility.

Corrective Measure: Improper installation will be corrected by the contractor. Failure of the roofing surface due to a manufacturer defect is the responsibility of the manufacturer whose products have specific specifications, standards, and warranty coverage.

6–4–19   **Observation:** Roof tiles are broken.

Performance Guideline: Roof tiles should not be broken within the warranty period unless caused by natural events.

Corrective Measure: <u>One time only</u> during the warranty period, broken tiles that are not the result of natural events will be replaced with closely matching tiles.

Chimney

6–5–1   **Observation:** A crack in a masonry chimney cap or crown causes leakage.

Performance Guideline: It is common for caps to crack from expansion and contraction. As a result, leaks may occur.

Corrective Measure: Contractor will repair leaks in new chimney cap or crown. Applying caulk or another sealant is an acceptable repair.

Discussion: Inevitable expansion and contraction at caps and crowns can result in leaks. After the warranty period, maintenance is the consumers responsibility.

6–5–2   **Observation:** New chimney flashing leaks.

Performance Guideline: New chimney flashing should not leak under normal conditions.

Corrective Measure: The contractor will repair leaks in new chimney flashing that are not caused by ice buildup or by the consumer's actions or negligence.

Discussion: The accumulation of ice and snow on the roof is a natural occurrence and cannot be prevented by the contractor.

Gutters and Downspouts

6–6–1   **Observation:** The gutter or downspout leaks.

Performance Guideline: Gutters and downspouts should not leak.

Corrective Measure: The contractor will repair leaks in gutters and downspouts. If the leaks are the results of debris, it is not the responsibility of the contractor to repair. Sealants are acceptable.

Discussion: Occasionally water may get between the gutter and fascia. Following the warranty period, this is a consumer maintenance issue as it frequently is caused by gutters that have debris built up inside.

6–6–2   **Observation:** The gutter overflows during a heavy rain.

Performance Guideline: Gutters should not overflow during normal rain.

Corrective Measure: The contractor will repair the gutter if it overflows during normal rains.

Discussion: Gutters may overflow during a heavy rain. The consumer is responsible for keeping gutters and downspouts free from debris that could cause overflow.

6–6–3   **Observation:** Rainwater on the roof is not directed into the gutters.

Performance Guideline: Most of the rainwater should be diverted into gutters and downspouts except under unusual circumstances.

Corrective Measures: Contractor will adjust drip edges and other flashings to deflect water into the gutters.

6–6–4   **Observation:** Water remains in the gutter after a rain.

Performance Guideline: The water level should not exceed ½ inch in depth if the gutter is unobstructed by ice, snow, or debris.

Corrective Measure: The contractor will repair the gutter to meet the performance guideline.

Discussion: The consumer is responsible for maintaining gutters and downspouts and keeping them unobstructed. Contractors install residential gutters with minimal slope in order to maintain an attractive appearance. Installing gutters with a ⅟₃₂-inch drop per every 1 foot generally will prevent water from standing in the gutters. Even so, small amounts of water may remain in some sections of the gutter for a time after a rain. In areas with heavy rainfall and/or ice buildup, a steeper pitch or additional downspouts may be desirable. During fall season when leaves fall from trees, frequent removal of leaves may be necessary, perhaps weekly.

Skylights and Light Tubes

6–7–1   **Observation:** A skylight or a light tube leaks.

Performance Guideline: Skylights and light tubes should be installed in accordance with the manufacturer's installation instructions. Leaks resulting from improper installation are considered excessive.

Corrective Measure: The contractor will repair any improperly installed skylight and light tube to meet the performance guideline.

Discussion: Condensation on interior surfaces is not a leak.

7

Plumbing

Note: Remodeling Specific Guideline: *Where applicable, in the following guidelines the contractor is responsible only for areas of the home where work was performed as specified in the contract, and not for the entire home.*

Water Supply System

7–1–1   **Observation:** A pipe, valve, or fitting leaks.

Performance Guideline: No leaks of any kind should exist from any water pipe, valve, or fitting.

Corrective Measure: The contractor will make repairs to eliminate leaks.

7–1–2   **Observation:** Condensation is observed on pipes, fixtures, or plumbing supply lines.

Performance Guideline: Condensation on pipes, fixtures, and plumbing supply lines may occur at certain temperatures and indoor humidity levels.

Corrective Measure: The consumer is responsible for controlling humidity in the home. No corrective action is required by the contractor.

Discussion: The consumer may insulate pipes and supply lines.

7–1–3   **Observation:** A water pipe freezes or has burst due to freezing temperatures.

Performance Guideline: The contractor should provide adequate freeze protection to drain waste, vent, and supply line in accordance with the prevailing building code.

Corrective Measure: The contractor shall repair bursted pipe and bring freeze protection measure into compliance with the prevailing building code.

Discussion: The consumer is responsible for maintaining the home at appropriate temperatures and taking appropriate precautions during freezing conditions (e.g. leaving faucets dripping and cabinet doors open).

7–1–4   **Observation:** The water supply system fails to deliver water.

Performance Guideline: All on-site service connections to the municipal water main or private water supply are the contractor's responsibility at the time of substantial completion.

Corrective Measure: The contractor will repair the water supply system if the failure results from improper installation or failure of materials and if the connections are a part of the construction agreement. Conditions beyond the control of the contractor that disrupt or eliminate the water supply are not covered.

7–1–5   **Observation:** The water pressure is low.

Performance Guideline: The plumbing system, including wells provided by contractor, shall be designed and installed in accordance with the prevailing plumbing code. The system should deliver water at the expected water pressure based on the pressure supplied to the home. If the water source is from a public source or (a well) supplied by the consumer, then the contractor has no control over the pressure thus no responsibility.

> **Remodeling Specific Guideline:** If the problem is a condition caused by the existing system, the consumer is responsible for necessary corrections (e.g., sludge, pipe corrosion, scaling, etc.).

Corrective Measure: When water pressure is determined by public or private (well) water source, no corrective action is required by contractor.

Discussion: Low water flow may result from the installation of low-flow fixtures required by the prevailing plumbing code. These fixtures affect flow, not water pressure.

7–1–6   **Observation:** The water pressure is high.

Performance Guideline: The plumbing system, including wells provided by contractors, shall be designed and installed according to the prevailing plumbing code.

Corrective Measure: If the water pressure exceeds what is specified in the prevailing plumbing code, the contractor shall install a water pressure reducing device to adjust pressure to acceptable pressure according to prevailing code.

7–1–7   **Observation:** A water supply line is noisy.

Performance Guideline: Because of the flow of water and pipe expansion/contraction, the water piping system may emit some noise.

Corrective Measure: The contractor cannot eliminate all noises caused by water flow and pipe expansion/contraction.

Discussion: Banging sounds can be caused by high-pressure dishwashers, washing machines, and high-efficiency toilets.

7–1–8   **Observation:** The water heater is not properly secured.

Performance Guideline: The water heater should be properly installed per manufacture's specifications and the prevailing building code.

Corrective Measure: The contractor will secure the water heater as necessary to meet the performance guideline.

7–1–9   **Observation:** Water hammer is evident when a fixture or valve is turned off.

Performance Guideline: Pipes should not make a pounding noise called water hammer.

Remodeling Specific Guideline: If the problem is a condition caused by the existing system, the consumer is responsible for necessary corrections. Some newer fixtures are more likely to present water hammer due to closing suddenness of the water flow.

Corrective Measure: The contractor will provide the water hammer prevention required by the prevailing plumbing code.

Plumbing Fixtures

7–2–1   **Observation:** A faucet leaks.

Performance Guideline: A faucet should not leak.

Corrective Measure: The contractor will stop the leak or replace a leaking faucet if the contractor provided the fixture.

7–2–2 **Observation:** Water flows outside a bathtub or shower.

Performance Guideline: Bathtubs and showers should be installed properly according to the manufacturer's guidelines.

Corrective Measure: The contractor will repair bathtub or shower leaks as necessary to meet the performance guideline during the warranty period.

Discussion: Proper repair can be achieved by sealing areas around bathtubs and showers. The consumer is responsible for maintaining caulk seals after the point of substantial completion of the project. The consumer is responsible for leaks related to the use of curtains in bathtubs and showers.

7–2–3 **Observation:** A plumbing fixture, appliance, or trim fitting is defective.

Performance Guideline: Plumbing fixtures, appliances, and trim fittings should perform in accordance with manufacturer's specifications at the time of substantial completion of the project.

Corrective Measure: Defective fixtures, appliances, or trim fittings are covered under the manufacturer's warranty. Contractor will be responsible for repairing or replacing fixtures, appliances and trim fittings provided by contractor during the warranty period.

7–2–4 **Observation:** The surface of a plumbing fixture is cracked, chipped, or scratched.

Performance Guideline: Cracks, chips, or scratches in surfaces of showers, bathtubs, and sinks are considered excessive if they are visible from 3 feet in normal lighting conditions at the time of substantial completion of the project.

Corrective Measure: The contractor shall repair any fixture that does not meet the performance guideline. The contractor is not responsible for repairs unless the damage is reported to the contractor prior to substantial completion or move in.

7–2–5 **Observation:** The surface of a plumbing fixture is stained. The fixture has an accumulation of minerals on it, or the fixture has been etched or corroded.

Performance Guideline: High mineral content in water can cause staining or corrosion of plumbing fixtures.

Corrective Measure: No corrective action is required by the contractor.

7–2–6   **Observation:** A bathtub or shower enclosure base flexes excessively.

Performance Guideline: The bathtub or shower enclosure should be installed according to the manufacturer's instructions and perform in accordance with the manufacturer's specifications; however, some noticeable flex can be expected and is normal.

Corrective Measure: The contractor will repair the base to meet the manufacturer's guideline.

Discussion: It is normal for bathtub and shower enclosure designs and materials to exhibit some flexing. Minimal noises may be associated with such movement.

7–2–7   **Observation:** A vanity top with a one-piece sink top is cracked.

Performance Guideline: Vanity tops should not have cracks.

Corrective Measure: The contractor will repair or replace the vanity top to meet the performance guideline. Cracks must be noted prior to substantial completion of the project.

7–2–8   **Observation:** A plumbing fixture does not deliver hot water.

Performance Guideline: The plumbing lines and fixtures should be correctly installed.

Corrective Measure: The contractor will correct the plumbing lines and/or adjust fixtures to meet the performance guideline.

Discussion: Hot water tanks or tankless water heaters are sometimes set at low temperatures to conserve energy and prevent young children from burning themselves. Some building codes limit the maximum temperature.

7–2–9   **Observation:** Hot water takes too long to get to a fixture.

Performance Guideline: Plumbing system will deliver hot water to the fixture.

Corrective Measure: Contractor will repair to meet the performance guideline. If hot water is delivered to the fixture, no corrective action is required by the contractor.

Discussion: Delays in water reaching certain fixtures can occur due to the proximity of fixture from the hot water source. Longer delays are typical with tankless water heaters. Point of use heaters or recirculating pumps are an option that can be installed by the consumer.

7–2–10   **Observation:** The sink waste disposer is clogged.

Performance Guideline: Disposer should not clog during normal use.

Corrective Measure: The contractor will correct the disposer unless the clog is due to improper use.

Discussion: Consumer should be aware that certain food items are difficult to grind in a disposer and will cause a clog and back up in the drain (ie., potato peels, carrot peels, shrimp shells, and other fibrous foods).

Sanitary Sewer or Septic System

7–3–1   **Observation:** Wastewater drain system components and pipes clog frequently, drain slowly, or back up into sink.

Performance Guideline: Sewer components and drains should drain as designed.

Remodeling Specific Guideline: If the problem is caused by issues with drain lines contained in the existing system, the consumer is responsible for necessary corrections. In a remodeled home drain blockage or inadequate flow can occur due to tree roots, sludge, etc. … found in an existing drain line.

Corrective Measure: The contractor will correct problems caused by improper installation. If consumer action or negligence caused the problem, the consumer is responsible for the necessary repairs.

Discussion: There are many items that should not be introduced to the wastewater drains, such as:
• Grease
• Fat
• Fruit
• Vegetable peels, rinds, and other fibrous foods
• Feminine products
• Cotton swabs
• Diapers
• Baby wipes
• Dental floss
• Paper towels

7–3–2   **Observation:** The septic (onsite wastewater collection and treatment system) does not operate as designed.

Performance Guideline: The septic system will function as designed and approved by the applicable local governing authority.

Corrective Measure: The contractor will correct problems caused by improper installation. The consumer is responsible for the proper maintenance of the system. If consumer action is the cause, the consumer is responsible for correcting the problem.

Discussion: Septic systems are sensitive biological systems whose function is dependent upon proper biological conditions and chemical balance that will promote the livelihood of the bacteria living with the system. More information regarding proper operation and management can be found at the National Environmental Services Center at www.nesc.wvu.edu/subpages/septic.cfm. Consumer actions that will negate corrective measure by the contractor under this performance guideline include but are not limited to the following:

- Connection of sump pump, roof drains, or backwash from a water conditioner into the system.

- Placement of non-biodegradable or nominally biodegradable items, such as personal wipes and feminine hygiene products into the system.

- Excessive use of a food waste disposer.

- Placement of surfaces not permeable to water over the disposal area of the system.

- Allowing vehicles to drive or park over the disposal area of the system.

- Failure to pump out the septic tank periodically, as required.

- Use that exceeds the system's design standards.

- Lack of vegetation maintenance over drain fields.

- Allowing water to pond over the disposal area.

- Using the system during chemotherapy drug treatment (including use on a temporary basis).

7–3–3   **Observation:** A toilet does not discharge wastewater properly.

Performance Guideline: Toilet should perform in accordance with manufacturer's specification.

Corrective Measure: The contractor will repair or replace the toilet not meeting the performance guideline.

Discussion: Toilets are designed to flush personal waste and toilet paper. Consumers should not flush products that are nominally biodegradable, such as personal wipes, paper towels, or napkins. Similarly, non-biodegradable items like feminine hygiene products, cotton balls or plastics should not be flushed. While many of these items may clear the toilet, they may cause blockage in the drainpipe. Consumers should avoid flushing prescription medications and other drugs.

7–3–4   **Observation:** The flushing level does not result in initiating a flush or water constantly enters the toilet.

Performance Guideline: The flushing mechanism should function properly at the time of substantial completion.

Corrective Measure: The contractor shall repair the flushing mechanism so that it operates properly.

7–3–5   **Observation:** A sewer odor is noticeable inside the home coming from the wastewater system.

Performance Guideline: A sewer odor should not be detectable inside the home under normal conditions.

Corrective Measure: The contractor should take the steps necessary to meet the performance guideline.

Discussion: The consumer should keep the plumbing traps filled with water. Extended non-use of a water fixture can allow the water in its trap to evaporate, thus providing a path for sewer gases to enter the home. Depending on humidity conditions, the consumer should fill traps by adding a quart of water to bathtubs, laundry tubs, and the like that are not used regularly, approximately every couple of months.

7–3–6 **Observation:** The grinder pump stops working (alarm sounding).

Performance Guideline: A grinder pump should work as indicated by manufacturer unless the power is off.

Corrective Measure: The contractor should take the steps necessary to meet the performance guideline.

Discussion: Grinder pump breakers can trip due to a close lightning strike. The consumer is responsible to maintain the pump according to manufacturer's instructions. Grinder pumps will fail if items such as nominally biodegradable, such as personal wipes, paper towels, or napkins. Similarly, non-biodegradable items like feminine hygiene products, diapers, some wipes, cotton balls, or plastics. Grinder pumps that use floats to sense the level in the holding tank are prone to grease build up and may turn the pump on unnecessarily or not at all causing the tank to fill up with sewage and possibly back up in the home or yard.

Electrical

Note: **Remodeling Specific Guideline:** *Where applicable, in the following guidelines the contractor is responsible only for areas of the home where work was performed as specified in the contract and not for the entire home. Some existing circuitry protection outlets or switches may be inadequate for newer lighting, appliances, and modern equipment.*

Fuses and Circuit Breakers

8–1–1 **Observation:** A *ground fault circuit interrupter* (GFCI) or *arc fault circuit interrupter* (AFCI) trips frequently.

Performance Guideline: GFCIs and AFCIs should perform as intended and will be installed and tested in accordance with prevailing electrical codes during warranty period.

Corrective Measure: The contractor will install ground fault and arc fault circuit interrupters in accordance with the prevailing electrical codes. Tripping is to be expected; however, the contractor will repair or replace components that frequently trip due to component failure or incorrect installation during warranty period.

Discussion: AFCIs are installed to protect bedroom circuits and all other habitable areas of a residence. GFCIs protect outlets in wet areas (e.g., bathrooms, kitchens, garages, laundry, exterior, etc.) Because outlets protected by GFCIs may be connected in a series, it may not be readily apparent that an inoperative convenience outlet is the result of a tripped GFCI in another room (not necessarily in the electrical panel). Both ground fault and arc fault circuit interrupters are sensitive devices that detect ground fault and arc fault conditions and consumers occasionally will experience nuisance tripping. The most common causes of nuisance tripping by AFCIs are damaged cords or plugs on consumers' lamps, small appliances, or other devices. Some vacuum cleaners, exercise equipment, light fixtures, and electronics may trip an AFCI-protected circuit. Static electricity, some electronic devices, televisions, computers, and printers may also cause nuisance tripping of circuit interrupters. The consumer should pay particular attention to refrigerators and freezers, as nuisance tripping of these devices may result in food spoilage.

8–1–2   **Observation:** A *circuit breaker* trips.

Performance Guideline: Circuit breakers should not be tripped by normal usage.

Corrective Measure: The contractor will check wiring, circuits, and components for conformity with prevailing electrical code. The contractor will correct noncompliant elements during warranty period.

Discussion: Prevailing electrical code determines the types of circuits, their locations, and their design loads. When maximum capacity is exceeded the fuse or breaker will trip. If a 15-amp circuit is tripped, the consumer should try plugging the device into a 20-amp circuit to see if the amperage of the device exceeds the lower 15-amp capacity. Blown fuses and tripped breakers are symptoms of a problem in some part of the home's electrical system or a consumer product connected to the system. Although components may be defective, consumer-owned fixtures and appliances usually are responsible for electrical malfunctions and nuisance tripping, larger devices such as treadmills and ellipticals, may not operate properly on these circuits designed for lower loads (e.g., bedrooms). The consumer should unplug or disconnect fixtures and appliances on the circuit and then replace the fuse or reset the breaker. If the problem reoccurs, the consumer should notify the contractor.

8–1–3   **Observation:** The home has lost partial power.

Performance Guideline: The electrical supply to the home should be installed in accordance with the prevailing electrical codes.

Corrective Measure: The contractor will check the wiring for conformity with prevailing electrical codes. The contractor will correct noncompliant elements during the warranty period.

Discussion: The consumer should check the circuit breakers and reset as needed. If this does not correct the situation, one leg of the power supply serving the home may no longer be operational. This may happen following a storm. The consumer should contact the local utility company and report the situation. The contractor is not responsible for the failure of wiring or connectors located before the service enters the home or of the adequate supply of service by the local utility company.

Outlets and Fixtures

8–2–1   **Observation:** Electrical outlets, switches, or fixtures malfunction.

Performance Guideline: All electrical outlets, switches, and fixtures should operate as designed.

Remodeling Specific Guideline: Existing recess cans may malfunction that are not IC (insulation contact) rated. Over time the adjacent insulation required gap for heat dissipation maybe missing. The light will flicker, burn out, or turn off when thermal coupling cuts the power.

Corrective Measure: The contractor will repair or replace malfunctioning electrical outlets, switches, and fixtures to meet the performance guideline during the warranty period.

8–2–2   **Observation:** Wiring fails to carry its designed load.

Performance Guideline: Wiring should be capable of carrying the designed load for normal residential use.

Corrective Measure: The contractor will verify that wiring conforms to prevailing electrical code. The contractor will correct wiring that does not conform to meet the performance guideline during the warranty period.

Discussion: Consumer needs to be aware of the circuit load capacity for specific breakers in their home, and not exceed that capacity. If they exceed the capacity of the circuit that has been provided per code specification, it will trip.

8–2–3   **Observation:** Interior receptacle or switch covers protrude from the wall.

Performance Guideline: Interior receptacle or switch covers should not protrude more than $\frac{1}{16}$ inch from the wall.

Remodeling Specific Guideline: Covers installed on an existing wall do not fall under this guideline.

Corrective Measure: The contractor will adjust the covers to meet the performance guideline.

Discussion: Some textured walls or tile may not allow a cover to be installed flush.

8–2–4   **Observation:** The consumer's 220-volt appliance plug does not fit the outlet provided by the contractor.

Performance Guideline: The contractor should install electrical outlets required by the prevailing electrical code.

Remodeling Specific Guideline: New 220-volt electrical outlets will not accept an appliance designed to be with older 220-volt outlets. The consumer is responsible for obtaining the appropriate appliance plug.

Corrective Measure: No corrective action is required by the contractor.

Discussion: The consumer is responsible for obtaining an appliance plug that is appropriate for their appliance and fits the outlets provided by the contractor.

8–2–5   **Observation:** Lighting dims or flickers when other electrical devices are in use.

Performance Guideline: General lighting outlets should be installed per the prevailing electrical codes.

Corrective Measure: The contractor will repair or replace outlets to meet the performance guideline.

Discussion: Lighting outlets are designed for moderate use devices such as clocks, radios and lamps. When larger devices such as vacuum cleaners, copiers, space heaters and irons are added, the startup surge may cause lights to dim or flicker. Air conditioners and heat pumps have the potential to cause all light fixtures in the home to dim or flicker. All circuits have the potential for dimming.

8–2–6   **Observation:** Lighting fixture on a dimmer switch does not operate properly.

Performance Guideline: Dimmer switches should be installed that are compatible to the bulbs installed.

Corrective Measure: The contractor will repair or replace bulbs or switches to meet the performance guideline, if provided by the contractor.

Discussion: Light fixtures require specific bulbs specified by the manufacturer. Some bulbs, particularly LED, are not compatible with standard dimmers and will blink, hum, flicker or not dim properly.

8–2–7   **Observation:** Low-voltage lighting flickers.

Performance Guideline: Low-voltage lighting should operate as designed.

Corrective Measure: The contractor will repair or replace malfunctioning low-voltage fixtures to meet the performance guideline during the warranty period.

Discussion: Low-voltage transformers are sized according to the fixtures that are installed on a circuit. The transformers must have sufficient ventilation around them. Consumers who add fixtures or change the wattage in fixtures after the circuit is designed and installed may cause the transformer to be undersized for a particular application.

8–2–8   **Observation:** Ceiling fan vibrates excessively and/or is noisy.

Performance Guideline: The contractor should install ceiling fans in accordance with the manufacturer's instructions (including blade balances).

Corrective Measure: The contractor will correct any fan installation not in accordance with the performance guideline if the fan was supplied and installed by the contractor.

Discussion: There are varying levels of performance for ceiling fans and some noise or vibration may be inherent in the specific fan installed.

8–2–9   **Observation:** A smoke or carbon monoxide detector chirps or otherwise malfunctions.

Performance Guideline: Detectors should operate as designed at substantial completion of the project.

Corrective Measure: The contractor will repair or replace the smoke or carbon monoxide detector to meet the performance guideline during the warranty period.

Discussion: Most smoke or carbon monoxide detectors are powered by both the electrical power and a backup battery. Chirping may indicate intermittent power loss, power surges or, most typically, that the battery is weak or is not installed. If the chirping occurs on a new detector, the contractor will check the battery, verify that the detector is wired correctly, and replace the device if necessary. Safety officials recommend that consumers change the batteries in detectors semiannually when Daylight Saving Time begins and ends.

8–2–10   **Observation:** Telephone and cable television wiring do not operate as intended.

Performance Guideline: Telephone and cable television wiring should be installed by the building contractor in accordance with the prevailing building codes.

Corrective Measure: The contractor shall repair telephone or cable television wiring not meeting the performance guideline during the warranty period.

Discussion: The contractor is not responsible for the failure of wiring or connectors located before the service provider enters the home.

8–2–11   **Observation:** Recessed can lights turn off.

Performance Guideline: Recessed lighting should operate as designed.

Remodeling Specific Guidelines: Older recessed can lights were not rated for insulation contact "(IC)". If a consumer is adding insulation near existing recessed can lights not rated IC, proper baffling and spacing is required to avoid a fire hazard or bulbs burning out.

Corrective Measure: The contractor shall repair recessed can lights not meeting the performance guideline.

Discussion: Recessed can lights have temperature sensors, or "thermal cutouts" that shut the lights off if the temperature gets too high. The temperature will get too high if a bulb higher than the maximum design wattage is used. Lights fitted with an enclosed trim (shower trim) will shut off if left on for an extended time and are operating as intended.

9

Interior Climate Control

Note: Remodeling Specific Guideline: *Where applicable, in the following guidelines the contractor is responsible only for areas of the home where work was performed as specified in the contract, not for the entire home.*

Air Infiltration and Drafts

9–1–1   **Observation:** Air infiltrates around exterior doors or windows.

Performance Guideline: Weather stripping should be installed and sized properly to seal the exterior door when closed. Windows will be installed per the manufacturer's instructions.

Corrective Measure: The contractor will correct to meet the performance guideline. Some infiltration is usually noticeable around doors and windows.

Discussion: At times of high wind or temperature differentials inside and outside the home, there may be noticeable air movement around a closed door's perimeter or window. In high-wind areas, the consumer may elect to have storm windows and doors installed to further reduce drafts. Doors must have gaps at their perimeter to accommodate expansion and contraction due to variations in temperature and/or humidity and to enable the door to operate over a wide range of environmental conditions. Weather stripping seals the gaps required for proper operations to prevent excessive air infiltration. A small glimmer of light seen at the corners of the door unit is normal. Weather stripping should be kept clean and maintained by the consumer.

9–1–2   **Observation:** A draft comes through an electrical outlet.

Performance Guideline: Electrical outlets and switch boxes on exterior walls may allow air to flow through or around an outlet into a room, as allowed by prevailing building code.

Corrective Measure: No corrective action is required by the contractor, except to conform to the prevailing building code.

Discussion: To increase energy efficiency, the consumer may elect to install foam insulation pads under switch and outlet plates to help decrease drafts.

9–1–3   **Observation:** The temperature or humidity of the home seems to change when the central vacuum system is operated.

Performance Guideline: Temperature of home may change when central vacuum system is operated.

Corrective Measure: No corrective action is required by the contractor.

Discussion: Most central vacuum systems expel air to the outside (i.e., out of the home). This results in a partial vacuum that causes outside air to be drawn into the home to make up for the expelled air. The introduction of unconditioned outside air may be perceptible.

Humidity Control and Condensation

9–2–1   **Observation:** Water, ice, frost, or condensation is observed on the interior frame or glass surface of a window or exterior door.

Performance Guideline: Windows and exterior doors should be installed in accordance with the manufacturer's instructions and the prevailing building codes.

Corrective Measure: No corrective action is required by the contractor.

Discussion: Condensation usually results from conditions beyond the contractor's control. Moisture in the air can condense into water and collect on cold surfaces, particularly in the winter months when the outside temperature is low. Blinds and drapes can prevent air within the home from moving across the cold surface and picking up the moisture. Occasional condensation on windows and doors in the kitchen, bath, or laundry area is also common. It is the consumer's responsibility to maintain proper humidity by properly operating heating and cooling systems' exhaust fans and allowing moving air within the home to flow over the interior surface of the windows. In hot, humid climates, condensation can occur on the outside of windows when the outdoor humidity is especially high (in early mornings when windows are cool). Air conditioning vents are usually aimed at windows and glass doors to maximize comfort and can cause surface condensation.

Ducts and Airflow

9–3–1   **Observation:** The *ductwork* makes noises.

Performance Guideline: Ductwork should be constructed and installed in accordance with applicable mechanical code requirements.

Corrective Measure: No corrective action is required by the contractor unless the *duct* does not comply with the prevailing building code.

Discussion: Metal expands when it is heated or subjected to pressure during startup and contracts when it cools. The ticking or crackling sounds caused by the metal's movement are common.

9–3–2   **Observation:** The ductwork produces excessively loud noises commonly known as "oil canning."

Performance Guideline: The stiffening of the ductwork and the thickness of the metal used should be such that ducts do not "oil can." The booming noise caused by oil canning is considered excessive.

Corrective Measure: The contractor shall correct the ductwork to eliminate oil canning.

9–3–3   **Observation:** There is airflow noise at a *register.*

Performance Guideline: The register should be correctly installed according to the prevailing building code.

Corrective Measure: No corrective action is required by the contractor unless registers are not installed according to the prevailing building code.

Discussion: Under certain conditions, there will be some noise with the normal flow of air even when registers are installed correctly.

9–3–4   **Observation:** The ductwork is separated or detached.

Performance Guideline: Ductwork should remain intact and securely fastened.

Corrective Measure: The contractor will reattach and secure all separated or unattached ductwork.

9–3–5   **Observation:** There is insufficient airflow to registers.

Performance Guideline: The ductwork should be correctly installed according to the prevailing building code and the applicable mechanical code.

Corrective Measure: The contractor will correct ductwork to meet the performance guideline. If the airflow is adequate to properly condition the room, no corrective action is required of the contractor.

Discussion: The adequacy of airflow may be subjective. See Sections 9–4–1 and 9–4–4 regarding the adequacy of the heating and cooling systems.

9–3–6   **Observation:** There is moisture accumulating on supply registers.

Performance Guideline: The ductwork should be correctly installed according to the prevailing building code and the applicable mechanical code.

Corrective Measure: No corrective action is required by the contractor unless registers are not installed according to the prevailing building code.

Discussion: Condensation usually results from conditions beyond the contractor's control. Moisture in the air can condense into water and collect on cold surfaces.

Heating and Cooling Systems

9–4–1   **Observation:** The heating system is inadequate.

Performance Guideline: The heating system should be capable of producing an inside temperature of 70 degrees Fahrenheit, as measured in the center of each room at a height of 5 feet above the floor under local outdoor winter design conditions. National, state, or local energy codes supersede this performance guideline where such codes have been adopted. Work should be done in accordance with the prevailing building codes.

> **Remodeling Specific Guideline:** For new living spaces created by remodeling projects, heating guidelines may not apply to areas where living space has been created without providing additional heating and/or resizing the ductwork.

Corrective Measure: The contractor will correct the heating system to provide the required temperature in accordance with the performance guideline or applicable code requirements. However, the consumer will be responsible for balancing dampers and registers and for making other necessary minor adjustments.

Discussion: Closed interior doors, closed registers, and dirty filters can restrict airflow and may affect the system's performance. To promote energy conservation, heating design codes typically are meant to maintain indoor temperature when the outdoor temperatures are within 10 degrees of normal, by climate zone. If the outdoor temperature is below that, the properly designed system will be unable to maintain 70 degrees Fahrenheit.

9–4–2   **Observation:** Some rooms are colder or hotter, or more humid than others.

Performance Guideline: The conditioning system should perform in accordance with the prevailing building code.

Corrective Measure: The contractor shall correct the flow of air to rooms to bring the flow into accordance with the prevailing building code.

Discussion: A temperature difference of several degrees Fahrenheit can be expected between rooms due to a number of factors, including registers that have been partially or completely closed, the number of people in a room (even sleeping), the number of appliances, even those that are ostensibly off, the amount of glass in the room, the number of and extent of exterior walls, the sun exposure at the time, and the temperature difference between inside and outside. When the temperature difference is relatively low in hot, humid climates, humidity can increase to give the perception of an increase in temperature.

9–4–3   **Observation:** The *radiant* floor has cold spots.

Performance Guideline: The radiant floor should be installed according to the manufacturer's instructions.

Corrective Measure: The contractor will correct to meet the performance guideline.

Discussion: Depending on the size, shape, flooring material, manufacturer, and type of radiant floor system, the number and size of cold spots in a floor will vary. A normally operating radiant floor system may include cold spots in perimeter areas and in areas between the heating sources.

9–4–4 **Observation:** The cooling of a room is inadequate.

Performance Guideline: If air conditioning is installed by the contractor, the cooling system should be capable of maintaining a temperature of 78 degrees Fahrenheit, as measured in the center of each room at a height of 5 feet above the floor under local outdoor summer design conditions. National, state, or local codes will supersede this guideline where such codes have been adopted. Work should be done in accordance with the prevailing building codes.

> **Remodeling Specific Guideline:** For new living spaces created by remodeling jobs, cooling guidelines may not apply to areas where living space has been created without providing additional cooling and/ or resizing the ductwork.

Corrective Measure: The contractor will correct the cooling system to provide the required temperature in accordance with the applicable code requirements.

Discussion: Closed interior doors without proper gap at bottom of the door, closed registers, and dirty filters can restrict airflow and may affect the system's performance. Some cooling systems have two filters. To promote energy conservation, cooling design codes typically are meant to maintain indoor temperature when the outdoor temperatures are within 15 degrees of normal. If the outdoor temperature is above normal, based on climate zone, the properly designed system will be unable to maintain 78 degrees Fahrenheit.

9–4–5 **Observation:** The air handler or furnace vibrates.

Performance Guideline: The units should be installed in accordance with the manufacturer's instructions and the prevailing building codes.

Corrective Measure: The contractor will correct the items according to the manufacturer's instructions and prevailing building code requirements.

Discussion: Under certain conditions, some vibration may occur with the normal flow of air when air handlers and furnaces are installed correctly. Debris in the furnace or air handler could cause the unit to become out of balance and vibrate. It is the consumer's responsibility to keep units clear of debris.

9–4–6   **Observation:** A condensate line is clogged.

Performance Guideline: Condensate lines should be free of all clogs at the time of substantial completion and have consistent pitch to drain as designed.

Corrective Measure: The contractor shall correct clogs existent before substantial completion. If a clog occurs after substantial completion of the project, no corrective action is required of the contractor.

Discussion: Condensate lines will eventually clog under normal use. The consumer is responsible for checking and maintaining clear lines. The addition of a tablespoon of bleach at the condensate trap can inhibit the growth of algae that is the frequent cause of clogging.

9–4–7   **Observation:** A leak in a refrigerant line or fittings.

Performance Guideline: Refrigerant lines and fittings should not leak.

Corrective Measure: The contractor will repair leaking refrigerant lines or fittings and recharge the air conditioning/heat pump unit unless the damage was caused by the consumer's actions or negligence.

9–4–8   **Observation:** There is condensation on the outside of air handlers, refrigerant lines, or ducts.

Performance Guideline: Moisture can be expected to condense and/or freeze on the exterior surfaces of air handlers, lines, and ducts when the air temperature is different from the surface temperature.

Corrective Measure: No corrective action is required by the contractor unless the condensation is directly attributed to faulty installation.

Discussion: Condensation is most likely to occur when air handlers, refrigerant lines, or ducts are in unconditioned locations such as a crawl space, basement, attic, or in exterior locations. Condensation usually results from conditions beyond the contractor's control. Moisture in the air can condense to form water and collect on cold duct surfaces, particularly in the summer months when the humidity is high.

Ventilation

9–5–1   **Observation:** Kitchen or bath fans allow air infiltration or make a flapping noise.

Performance Guideline: Bath and kitchen fans should be installed in accordance with the manufacturer's instructions and prevailing building code requirements and perform in accordance with the manufacturer's specifications.

Corrective Measure: No corrective action is required by the contractor if the fan installation meets the performance guideline.

Discussion: It is possible for outside air to enter the home through a ventilation fan. The dampers in most fans do not seal tightly. It is possible for the damper to be lodged open due to animal activity (including nesting in the outside opening), or the accumulation of grease, lint, and other debris. Maintenance of ventilating fans is the consumer's responsibility, and the consumer should make periodic inspections to assure the proper flow of air. Opening and closing exterior doors changes the interior pressure and may cause the dampers to open and close causing a noise.

9–5–2   **Observation:** HVAC vent or register covers protrude from a smooth wall or ceiling surface.

Performance Guideline: Registers should not protrude more than $\frac{1}{16}$ inch from a smooth wall or ceiling surface at the time of substantial completion of the project.

Corrective Measure: The contractor will correct to meet the performance guideline.

Discussion: Registers and vents may deflect over time. This can result in gaps occurring between the vents or register and the wall or ceiling. If the vent or register is securely attached, this is not a warranty item.

9–5–3   **Observation:** HVAC vent or register covers protrude from a rough or texture wall or ceiling surface.

Performance Guideline: Registers should not protrude more than 1/16 inch from a rough or texture wall or ceiling surface at the time of substantial completion of the project. Some texture wall finishes may not allow a register to be installed flush.

Corrective Measure: The contractor will correct to meet the performance guideline.

Discussion: Registers and vents may deflect over time. This can result in gaps occurring between the vents or register and the wall or ceiling. If the vent or register is securely attached, this is not a warranty item.

9–5–4   **Observation:** HVAC vent or register covers protrude from floors.

Performance Guideline: Registers should not protrude more than $\frac{1}{16}$ inch from floor surface at the time of substantial completion of the project.

Corrective Measure: The contractor will correct to meet the performance guideline or prevailing building code.

Discussion: Registers and vents may deflect over time. This can result in gaps occurring between the vents or register and the wall or ceiling. If the vent or register is securely attached, this is not a warranty item.

9–5–5   **Observation:** Exhaust fan does not discharge directly to the exterior.

Performance Guideline: Exhaust fans should vent in accordance with the prevailing building code.

Corrective Measure: The contractor will correct to meet the performance guideline.

Interior Finish

Note: Remodeling Specific Guideline: *Where applicable, in the following guidelines the contractor is responsible only for areas of the home where work was performed as specified in the contract, not for the entire home.*

Interior Doors

10–1–1   **Observation:** An interior door is warped.

Performance Guideline: Except as noted, interior doors should not become inoperable due to warping. A ¼ inch tolerance, as measured diagonally from corner to corner, is acceptable.

Corrective Measure: The contractor will correct or replace and refinish defective doors to match existing doors as closely as practical.

Discussion: In bathroom or utility areas, exhaust fans or an open window must be used to minimize moisture to prevent warpage of door units. The contractor is not responsible for refinishing if doors were finished by the consumer.

10–1–2   **Observation:** Bifold and bypass doors come off their tracks during normal operation.

Performance Guideline: *Bifold* and *bypass* doors should slide properly on their tracks.

Corrective Measure: One time only during the warranty period, the contractor will adjust the bifold or bypass doors that do not stay on its track or slide properly during normal operation.

Discussion: Proper operation should be verified by the consumer and the contractor at the time of substantial completion of the project and confrm that floor guides are in place.

Consumers should be aware that bifold and bypass doors are inherently more sensitive than swing doors and need to be treated accordingly. The consumer is responsible for cleaning and maintenance necessary to preserve proper operation.

10–1–3   **Observation:** Barn doors or pocket doors roll open on their own or do not stay closed.

Performance Guideline: Barn doors and pocket doors should not roll open or shut on their own.

Corrective Measure: <u>One time only</u> during the warranty period, the contractor will adjust the barn or pocket door that does not stay in place during normal operation.

Discussion: Proper operation should be verified by the consumer and the contractor at the time of substantial completion of the project and that jam guides are properly installed.

10–1–4   **Observation:** A *pocket* door rubs in its pocket during normal operation.

Performance Guideline: Pocket doors should operate smoothly during normal operation.

Corrective Measure: <u>One time only</u> during the warranty period, the contractor will adjust the pocket door to meet the performance guideline.

Discussion: Pocket doors commonly rub, stick, or derail because of the inherent nature of the product. It is necessary for the door to also rub against the guides provided by the manufacturer.

10–1–5   **Observation:** A wooden door panel has shrunk or split.

Performance Guideline: Wooden door panels should not split to the point that light is visible through the door.

Corrective Measure: <u>One time only</u> during the warranty period, the contractor will fill splits in the door panel with wood filler and will match the paint or stain as closely as practical.

Discussion: Contractor may check if defective door is covered by a manufacturers' warranty.

10–1–6   **Observation:** A door rubs on *jambs* or contractor-installed floor covering.

Performance Guideline: Doors should not rub on jambs or contractor-installed floor covering.

Corrective Measure: <u>One time only</u> during the warranty, the contractor will adjust the door as necessary to meet the performance guideline.

10–1–7   **Observation:** A door edge is not parallel to the door jamb.

Performance Guideline: When the contractor installs the door frame and door, the door edge should be within ³⁄₁₆ inch of parallel to the door jamb.

Remodeling Specific Guideline: Where the contractor installs the door in an existing frame the performance guideline does not apply.

Corrective Measure: <u>One time only</u> during the warranty period, the contractor will adjust the door as necessary to meet the performance guideline.

10–1–8   **Observation:** A door swings open or closed from the force of gravity.

Performance Guideline: Doors should not swing open or closed from the force of gravity alone.

Remodeling Specific Guideline: This guideline does not apply where a door is installed in an existing wall that is out of plumb.

Corrective Measure: <u>One time only</u> during the warranty period, the contractor will adjust the door as necessary to meet the performance guideline.

10–1–9   **Observation:** A door hinge squeaks.

Performance Guideline: Door hinges should not squeak.

Corrective Measure: <u>One time only</u> during the warranty period, the contractor will adjust the door as necessary to meet the performance guideline.

10–1–10   **Observation:** Interior doors do not operate smoothly.

Performance Guideline: Doors should move smoothly with limited resistance.

Corrective Measure: <u>One time only</u> during the warranty period, the contractor will adjust the door to meet the performance guideline.

10–1–11   **Observation:** A doorknob or latch does not operate smoothly.

Performance Guideline: A doorknob or latch should not stick or bind during operation.

Corrective Measure: <u>One time only</u> during the warranty period, the contractor will adjust, repair, or replace knobs or latches that are not operating smoothly.

Discussion: Because locksets are rather complex mechanical devices, some may have a heavy or stiff feel but are operating as intended by the manufacturer. This can be true for locksets of all price ranges. Slamming doors or hanging items on the doorknob will affect knob or latch operation; it is not the contractor's responsibility to adjust or repair problems caused by such conditions.

Interior Stairs

10–2–1   **Observation:** An interior stair tread deflects.

Performance Guideline: The maximum vertical deflection of an interior stair tread should not exceed ⅛ inch at 200 pounds of force.

Corrective Measure: The contractor will repair the stair to meet the performance guideline.

10–2–2   **Observation:** Gaps exist between interior stair risers, treads, and/or skirts.

Performance Guideline: Gaps between adjoining parts that are designed to meet flush should not exceed ⅛ inch in width.

Corrective Measure: The contractor will repair or replace the parts as necessary to meet the performance guideline.

Discussion: The use of filler is an appropriate method to fill gaps.

10–2–3   **Observation:** A stair riser or tread squeaks.

Performance Guideline: Loud squeaks, pops or creeks caused by a loose stair riser or tread are considered excessive; however, totally squeakproof stair risers or treads cannot be guaranteed.

Corrective Measure: The contractor will refasten any loose risers or treads or take other reasonable and cost-effective corrective action, based on his or her best judgment, to reduce squeaking without removing treads or ceiling finishes.

Discussion: Squeaks in risers or treads may occur when a riser has come loose from the tread, deflects from the weight of a person and rubs against the nails that hold it in place. Movement may occur between the riser and the tread or other stairway members when one tread is deflected while the other members remain stationary. Using trim screws to fasten the tread to the riser from above sometimes will reduce squeaking. If there is no ceiling below, gluing or re-nailing the riser to the tread or shimming will reduce squeaks but completely eliminating squeaks is not always possible.

10–2–4   **Observation:** Gaps exist between interior stair railing parts.

Performance Guideline: Gaps between interior stair railing parts should not exceed ⅛ inch in width.

Corrective Measure: One time only during the warranty period, the contractor will ensure that individual parts of the railing are securely mounted. Any remaining gaps will be filled or the parts will be replaced to meet the performance guideline.

10–2–5   **Observation:** An interior stair railing lacks rigidity.

Performance Guideline: Interior stair railings should be installed in accordance with prevailing building codes.

Corrective Measure: The contractor will secure any stair railing parts that loosen with normal use, to meet the performance guideline.

Discussion: Stair railings are designed to protect an individual while stepping up and down a stairwell. Damages caused by the consumer from pulling, swinging, hanging, or sliding on railings may loosen the rail system and the contractor is not responsible for repair of such.

10–2–6   **Observation:** An interior balcony or horizontal railing lacks rigidity.

Performance Guideline: Interior railings should be installed in accordance with prevailing building codes.

Corrective Measure: The contractor will secure any railing parts that loosen with normal use, to meet the performance guideline.

Trim and Moldings

10–3–1   **Observation:** There are gaps at non-*mitered* trim and molding joints.

Performance Guideline: At the time of substantial completion of the project, openings at joints in trim and moldings, and at joints between moldings and adjacent surfaces, should not exceed 1/8 inch in width.

Corrective Measure: The contractor will repair joints to meet the performance guideline.

Discussion: Failing to control indoor relative humidity may cause separation of trim and moldings in excess of the performance guideline. Joints that separate under these conditions are not considered defective. The consumer is responsible for controlling temperature and humidity in the home.

10–3–2   **Observation:** Nails are not properly set or nail holes are not properly filled.

Performance Guideline: *Setting* nails and filling nail holes are considered part of painting and finishing. After finishing, nails and nail holes should not be readily visible from a standing position facing the surface at distance of 6 feet under normal lighting conditions. After staining, putty colors will not exactly match variations in wood color.

Corrective Measure: Where the contractor is responsible for painting, the contractor will take action necessary to meet the performance guideline. Puttying of nail holes in base and trim molding installed in unfinished rooms and areas not exposed to view (such as inside of closets) is not included in this guideline.

10–3–3   **Observation:** An inside corner is not coped or mitered.

Performance Guideline: Trim and molding edges at inside corners should be coped or mitered. However, square-edge trim and molding may be butted.

Corrective Measure: The contractor will finish inside corners to meet the performance guideline.

10–3–4   **Observation:** Trim or molding mitered edges do not meet.

Performance Guideline: At the time of substantial completion of the project, gaps between mitered edges in trim and molding should not exceed ⅛ inch.

Corrective Measure: The contractor will repair gaps that do not meet the performance guideline. Caulking or puttying with materials compatible with the finish is acceptable.

Discussion: Separation of trim and moldings in excess of the performance guideline may be caused by lack of control of indoor relative humidity. Joints that separate under these conditions are not considered defective. It is the consumer's responsibility to control temperature and humidity in the home.

10–3–5   **Observation:** Interior trim is split.

Performance Guideline: Splits, cracks, and checking greater than ⅛ inch in width are considered excessive.

Corrective Measure: One time only during the warranty period, the contractor will repair the affected area to meet the performance guideline. Refinished or replaced areas may not match surrounding surfaces exactly.

10–3–6   **Observation:** Hammer marks are visible on interior trim.

Performance Guideline: Hammer marks on interior trim should not be readily visible from a standing position facing the surface at a distance of 6 feet under normal lighting conditions.

Corrective Measure: The contractor will fill hammer marks and refinish or replace affected trim to meet the performance guideline. Refinished or replaced areas may not match surrounding surfaces exactly.

Discussion: Dents and marks on trim due to consumer's actions are not the contractor's responsibility.

10–3–7   **Observation:** Wood trim appearance is uneven.

Performance Guideline: Variations in natural wood trim are common.

Corrective Measure: No corrective action is required by the contractor.

Cabinets

10–4–1   **Observation:** Cabinets do not meet the ceiling or walls.

Performance Guideline: Gaps greater than ¼ inch in width are considered excessive.

Corrective Measure: The contractor will repair the gap with caulk, putty, scribe molding, or will reposition/reinstall cabinets to meet the performance guideline.

Discussion: Remodeling Specific Guideline: When installed in rooms with out-of-plumb walls or out-of-level floors and ceilings, "square" cabinets cannot be installed parallel to walls and ceilings and still be kept on line. For example, if the floor is not level and the installer measures up from it, snaps a line on which to place the tops of the wall cabinets, and then plumbs the first cabinet, one corner of the cabinet will leave the line, and the bottom of successive cabinets will not be in line. Similarly, cabinets will not line up with each other if they are installed on a level line, starting against an out-of-plumb wall instead of a plumb wall. The contractor should explain the aesthetic options to the consumer and select the best option with the consumer.

10–4–2   **Observation:** Cabinets do not line up with each other.

Performance Guideline: Cabinet faces more than ⅛ inch out of line, and cabinet corners more than ³⁄₁₆ inch out of line are considered excessive.

Remodeling Specific Guideline: The consumer and the contractor may agree to disregard this guideline to match or otherwise compensate for preexisting conditions.

Corrective Measure: The contractor will make necessary adjustments to meet the performance guideline.

Discussion: Remodeling Specific Guideline: In remodeling projects, many times the rooms are out of square, walls are not plumb, and floors are not level. Cabinets and countertops may have to be shimmed or otherwise adjusted to make the cabinets and countertops fit together properly. Cabinets may not fit flush against the walls on the ends or bottoms and may not fit flat against the floor.

10–4–3   **Observation:** A cabinet door or drawer front is warped.

Performance Guideline: Door or drawer warpage should not exceed ¼ inch as measured from the face frame to the point of furthermost warpage, with the door or drawer front in closed position.

Corrective Measure: The contractor will correct or replace doors and drawer fronts as necessary to meet the performance guideline.

Discussion: Failing to control indoor relative humidity may cause warpage that exceeds the performance guideline. Doors or drawers that warp under these conditions are not considered defective. It is the consumer's responsibility to control temperature and humidity in the home.

10–4–4   **Observation:** A cabinet door or drawer binds.

Performance Guideline: Cabinet doors and drawers should open and close with reasonable ease.

Corrective Measure: The contractor will adjust or replace cabinet door hinges and/or drawer hardware as necessary to meet the performance guideline.

10–4–5   **Observation:** A cabinet door will not stay closed.

Performance Guideline: The catches or closing hardware for cabinet doors should be adequate to hold the doors in a closed position.

Corrective Measure: One time only during the warranty period, the contractor will adjust or replace the door catches or closing hardware as necessary to meet the performance guideline.

10–4–6   **Observation:** Cabinet doors or drawer fronts are cracked.

Performance Guideline: Cabinet doors and drawer fronts should not crack.

Corrective Measure: The contractor will replace or repair cracked panels and drawer fronts. No corrective action is required by the contractor if the cracked drawer fronts or panels result from the consumer's abuse.

Discussion: Paint or stain on the repaired or replaced door or drawer front may not match the stain on the existing panels or drawer fronts. Grain patterns or intensity cannot be matched perfectly. The contractor will use his or her best efforts to match as closely as possible the stain on the existing panels or drawer fronts. However, some species of wood will age and darken over time and an exact match may not be possible. Use of manufacturer-provided touch-up kits is acceptable to address minor imperfections in the cabinet finish.

10–4–7 **Observation:** Cabinet units are not level.

Performance Guideline: Individual cabinets should not have a deviation of more than ³⁄₁₆ inch out of level.

> **Remodeling Specific Guideline:** The consumer and the contractor may agree to disregard this guideline to match or otherwise compensate for preexisting conditions.

Corrective Measure: The contractor will level cabinets to meet the performance guideline.

Discussion: Remodeling Specific Guideline: In remodeling projects, many times the rooms are out of square, walls are not plumb, and floors are not level. Cabinets and countertops may have to be shimmed or otherwise adjusted to make the cabinets and countertops fit together properly. Cabinets may not fit flush against the walls on the ends or bottoms and may not fit flat against the floor.

10–4–8 **Observation:** A cabinet frame is out of square.

Performance Guideline: A cabinet frame, when measured diagonally from corner to corner, should not exceed a difference of more than ¼ inch.

Corrective Measure: The contractor will repair or replace the cabinet to meet the performance guideline.

10–4–9 **Observation:** Cabinet doors do not align when closed.

Performance Guideline: Gaps between doors should not deviate more than ⅛ inch from top to bottom.

Corrective Measure: The contractor will adjust doors to meet the performance guideline.

10–4–10 **Observation:** Soft or self-closing hinges or drawer slides are not functioning properly.

Performance Guideline: Hinges and slides should operate properly.

Corrective Measure: One time only during the warranty period, the contractor will adjust hinges or slides to meet the performance guidelines.

10–4–11 **Observation:** Cabinet doors and drawer fronts delaminate or discolor.

Performance Guideline: Cabinet doors and drawer fronts should not delaminate or discolor.

Corrective Measure: Cabinet defects are covered by the manufacturer's warranty.

Discussion: Under counter steam and heat from appliances such as dishwashers, drawer microwaves, ovens, steam dryers, and others can also cause damage to cabinets and is not a contractor responsibility.

10–4–12 **Observation:** Cabinet shelves are sagging or deflecting.

Performance Guideline: Cabinet shelves should not sag with normal storage loads.

Corrective Measure: Cabinet defects are covered by the manufacturer's warranty.

Discussion: Consumer is responsible for proper use and not exceeding maximum storage weight.

Countertops

10–5–1 **Observation:** High-pressure laminate on a countertop is delaminated.

Performance Guideline: Countertops fabricated with high-pressure laminate coverings should not delaminate.

Corrective Measure: The contractor will repair or replace delaminated coverings unless the *delamination* was caused by the consumer's misuse or negligence.

Discussion: Consumers should refrain from leaving any liquids near the countertop seams or allowing the surface to become excessively hot. Under counter steam and heat from appliances such as dishwashers, drawer microwaves, ovens, steam dryers, etc., can also cause damage to countertops and is not a contractor responsibility.

10–5–2 **Observation:** The surface of high-pressure laminate on a countertop is cracked or chipped.

Performance Guideline: At the time of substantial completion of the project, cracks or chips greater than a 1/16 inch are considered excessive.

Corrective Measure: The contractor will repair or replace cracked or chipped countertops to meet the performance guideline only if they are reported at the time of substantial completion of the project.

10–5–3 **Observation:** Countertops are visibly scratched.

Performance Guideline: At the time of substantial completion of the project, countertops should be free of scratches visible from 6 feet under normal lighting conditions.

Corrective Measure: The contractor will repair scratches in the countertop to meet the performance guideline.

Discussion: Minor imperfections and scratches will be more visible in dark, glossy tops.

10–5–4 **Observation:** A countertop is not level.

Performance Guideline: Countertops should be no more than 3/8 inch in 10 feet out of parallel with the floor.

> **Remodeling Specific Guideline:** For projects where the floor is out of level, the countertop may be installed proportionately out of level.

Corrective Measure: The contractor will make necessary adjustments to meet the performance guideline.

Discussion: Remodeling Specific Discussion: In remodeling projects, many times the rooms are out of square, walls are not plumb, and floors are not level. Cabinets and countertops may have to be shimmed or otherwise adjusted to make the cabinets and countertops fit together properly.

10–5–5 **Observation:** A tile countertop or backsplash has uneven grout lines.

Performance Guideline: Tile will be installed with grout lines as defined in the manufacturer's installation instructions.

Corrective Measure: The contractor will make corrections as necessary to bring the grout lines into compliance to meet the performance guideline.

Discussion: Different tiles require different widths of grout lines. Some tiles are designed to have varied-width grout lines. Irregularly sized tiles will also often result in uneven and variable grout width.

10–5–6   **Observation:** Tile countertop or backsplash grout lines are cracked.

Performance Guideline: Tile grout is a cement product and is subject to cracking. Cracks that result in loose tiles or gaps of $\frac{1}{16}$ inch are excessive.

Corrective Measure: <u>One time only</u> during the warranty period, the contractor will repair the grout lines of cracks that result in loose tiles or gaps of $\frac{1}{16}$ inch by adding grout, caulking, or replacing grout.

10–5–7   **Observation:** The surface of countertop or backsplash tile has excessive lippage from the adjoining tile.

Performance Guideline: Generally, lippage greater than $\frac{1}{16}$ inch is considered excessive, except for materials that are designed with an irregular thickness (such as handmade tile).

Corrective Measure: The contractor will repair or replace the tile to meet the performance guideline, except for materials that are designed with an irregular thickness (such as handmade tile).

Discussion: Different types of tiles may have varying tolerances.

10–5–8   **Observation:** A natural stone, or solid-surface countertop is cracked.

Performance Guideline: At the time of substantial completion of the project, cracks greater than $\frac{1}{32}$ inch in width are considered excessive.

Corrective Measure: If the crack is found to be a result of faulty installation or product, the contractor will repair or replace the countertop. Patching is an acceptable repair.

Discussion: Imperfections in natural stone are inherent and normal and do not require corrective measures.

10–5–9   **Observation:** A natural stone countertop is discolored due to water, oil, pans, cleaners, etc.

Performance Guideline: At the time of substantial completion countertops should be clear of discoloration and sealed.

Corrective Measure: No corrective action is required by the contractor.

Discussion: Countertops can discolor during normal use. Resealing countertops is a home owner maintenance item.

10–5–10   **Observation:** A natural stone or solid-surface countertop is chipped.

Performance Guideline: At the time of substantial completion of the project, chips greater than $\frac{1}{32}$ inch in width are considered excessive.

Corrective Measure: The contractor will repair or replace affected areas to meet the performance guidelines. The use of an appropriate filler is an acceptable repair.

10–5–11   **Observation:** A natural stone, or solid-surface countertop has visible seams.

Performance Guideline: Seams may be visible and especially noticeable with certain countertop materials and darker finishes.

Corrective Measure: No corrective action is required by the contractor.

10–5–12   **Observation:** A natural stone countertop has excessive lippage between sections.

Performance Guideline: Lippage greater than $\frac{1}{32}$ inch is considered excessive.

Corrective Measure: The contractor will repair or replace the countertop to meet the performance guideline.

10–5–13   **Observation:** A solid-surface or laminate countertop has a bubble, burn, stain, or other damage.

Performance Guideline: At the time of substantial completion of the project, solid-surface or laminate countertops should be free of bubbles, burns, or stains.

Corrective Measure: The contractor will repair or replace the countertop to meet the performance guideline.

Discussion: Solid-surface and laminate products may be subject to damage by hot surfaces placed on or near the product. The consumer is responsible for maintaining the countertop and protecting it from damage.

10–5–14   **Observation:** Manmade cultured marble top has hairline cracking around or near the drain.

Performance Guideline: At the time of substantial completion of the project, no visible cracks should be apparent to the naked eye. The countertop should withstand water temperatures of 130 degrees Fahrenheit without cracking.

Corrective Measure: The contractor will repair or replace the countertop to meet the performance guideline.

Discussion: Cultured marble tops are sensitive to rapid temperature changes, and may become thermally shocked. This process will cause cracking of the gel coat finish at or near the point of the temperature change. Water heater should be set at or below 130 degrees Fahrenheit. The contractor is not responsible for damage caused by thermal shocking.

10–5–15   **Observation:** Quartz countertop has chips or pits.

Performance Guideline: Quartz material can have minor chips or pits.

Corrective Measure: The contractor will repair the countertop with a material for filling quartz.

Interior Wall Finish

Lath and Plaster

10–6–1   **Observation:** Cracks are visible on a finished wall or ceiling.

Performance Guideline: Cracks should not exceed 1/16 inch in width.

Corrective Measure: <u>One time only</u> during the warranty period, the contractor will repair cracks exceeding 1/16 inch in width. The contractor will touch up paint on repaired areas if the contractor was responsible for the original interior painting. A perfect match between original and new paint cannot be expected and the contractor is not required to paint an entire wall or room.

Gypsum Wallboard or Drywall

10–6–2 **Observation:** Nail pops are visible on a finished wall or ceiling.

Performance Guideline: Nail pops are a defect only when there are signs of spackle compound cracking or falling away.

Corrective Measure: <u>One time only</u> during the warranty period, the contractor will repair such blemishes. The contractor will touch up paint on repaired areas if the contractor was responsible for the original interior painting. A perfect match between original and new paint cannot be expected, and the contractor is not required to paint an entire wall or room. The contractor is not required to repair defects that are covered by wall coverings and that, therefore, are not visible.

Discussion: When drywall has been placed on lumber surfaces subject to shrinkage and warpage and which are not perfectly level and plumb, problems may often occur through stress and strain placed on drywall during the stabilization of the lumber, which is inherent in the construction of the home. Due to the initial stabilization problem that exists with the new home, it is impossible to correct each defect as it occurs, and it is essentially useless to do so. The entire home will tend to stabilize itself. Correcting the drywall near the end of the warranty period provides the consumer with the best possible solution.

10–6–3 **Observation:** Blisters or other blemishes are visible on a finished wall or ceiling.

Performance Guideline: Any such blemishes that are readily visible from a standing position facing the surface at distance of 6 feet under normal lighting conditions are considered excessive.

Corrective Measure: <u>One time only</u> during the warranty period, the contractor will repair such blemishes. The contractor will touch up paint on repaired areas if the contractor was responsible for the original interior painting. A perfect match between original and new paint cannot be expected, and the contractor is not required to paint an entire wall or room. The contractor is not required to repair defects that are covered by wall coverings and that, therefore, are not visible.

Discussion: When drywall has been placed on lumber surfaces subject to shrinkage and warpage and which are not perfectly level and plumb, problems may often occur through stress and strain placed on drywall during the stabilization of the lumber, which is inherent in the construction of the home. Due to the initial stabilization problem that exists with the new home, it is impossible to correct each defect as it occurs, and it is essentially useless to do so. The entire home will tend to stabilize itself. Correcting the drywall near the end of the warranty period provides the consumer with the best possible solution.

10–6–4   **Observation:** Cracked or exposed corner bead, excess joint compound, trowel marks, or blisters in tape joints are observed on the drywall surface.

    Performance Guideline: Defects resulting in cracked or exposed corner bead, *trowel marks,* excess joint compound or blisters in tape are considered excessive.

    Corrective Measure: <u>One time only</u> during the warranty period, the contractor will repair the affected area of the wall to meet the performance guideline.

10–6–5   **Observation:** Joints protrude from the surface.

    Performance Guideline: Any joints that are readily visible from a standing position facing the surface at distance of 6 feet under normal lighting conditions are considered excessive.

    Corrective Measure: <u>One time only</u> during the warranty period, the contractor will repair affected areas.

    Discussion: Visible joints often occur in long walls, stairwells, ceilings, and areas of two-story homes where framing members have shrunk and caused the drywall to protrude.

10–6–6   **Observation:** Angular *drywall joints* are uneven.

    Performance Guideline: This is a common condition that occurs with randomly applied materials.

    Corrective Measure: No corrective action is required by the contractor.

10–6–7 **Observation:** The texture of drywall does not match.

Performance Guideline: Any variations that are readily visible from a standing position facing the surface at a distance of 6 feet under normal lighting conditions are considered excessive.

Corrective Measure: The contractor will repair the affected area to meet the performance guideline.

10–6–8 **Observation:** Drywall is cracked.

Performance Guideline: Drywall cracks greater than $\frac{1}{16}$ inch in width are considered excessive.

Corrective Measure: <u>One time only</u> during the warranty period, the contractor will repair cracks and touch up paint in affected areas. The texture and paint color may not exactly match the existing texture and paint color.

10–6–9 **Observation:** Sprayed or textured ceilings have uneven textures.

Performance Guideline: This is a common condition that occurs with randomly applied materials.

Corrective Measure: No corrective action is required by the contractor.

Paint, Stain, and Varnish

10–6–10 **Observation:** Interior paint does not cover the underlying surface.

Performance Guideline: The surface being painted should not show through new paint when viewed from a standing position facing the surface at distance of 6 feet under normal lighting conditions.

Corrective Measure: The contractor will recoat affected areas as necessary to meet the performance guideline as closely as practical.

Discussion: The amount and direction of sunlight can have a significant effect on how a surface appears. It is not unusual for the underlying surface to be visible in direct sunlight; no corrective action is required of the contractor in such instances.

10–6–11   **Observation:** An interior surface is spattered with paint.

Performance Guideline: Paint spatters should not be readily visible on walls, woodwork, floors, or other interior surfaces when viewed from a standing position facing the surface at distance of 6 feet under normal lighting conditions.

Corrective Measure: The contractor will remove paint spatters to meet the performance guideline.

10–6–12   **Observation:** Brush and roller marks show on interior painted surface.

Performance Guideline: Brush marks should not be readily visible on interior painted surfaces when viewed from a standing position facing the surface at a distance of 6 feet under normal lighting conditions.

Corrective Measure: The contractor will refinish as necessary to meet the performance guideline and match affected areas as closely as practical.

10–6–13   **Observation:** *Lap marks* show on interior painted or stained areas.

Performance Guideline: Lap marks should not be readily visible on interior painted or stained areas when viewed from a standing position facing the surface at distance of 6 feet under normal lighting conditions.

Corrective Measure: The contractor will refinish as necessary to meet the performance guideline and match affected areas as closely as practical.

10–6–14   **Observation:** Interior painting, staining, or refinishing of repair work does not match.

Performance Guideline: A perfect match between original and new paint cannot be expected. Repairs required under the performance guideline will be finished to match the immediate surrounding areas as closely as practical.

Corrective Measure: No corrective action is required by the contractor.

Discussion: Where the majority of the wall or ceiling area is affected, the area will be painted from break line to break line. The contractor is not required to paint an entire room.

10–6–15 Observation: Tannin from wood has bled through the paint on interior trim.

Performance Guideline: This is a common condition with natural materials such as wood.

Corrective Measure: No corrective action is required by the contractor.

Wallpaper and Vinyl Wall Coverings

10–6–16 Observation: The wall covering has peeled.

Performance Guideline: The wall covering should not peel.

Corrective Measure: The contractor will reattach or replace the loose wall covering if the contractor installed the covering and peeling is not due to consumer actions.

Discussion: Wallpaper applied in high moisture areas is exempt from this guideline because the problem results from conditions beyond the contractor's control.

10–6–17 Observation: Patterns in wall covering are mismatched.

Performance Guideline: Patterns in wall coverings should match. Irregularities in the patterns themselves are the manufacturer's responsibility.

Remodeling Specific Guideline: This guideline does not apply if material is installed on existing out-of-plumb walls or where trim is not square with corners.

Corrective Measure: The contractor will correct the wall covering to meet the performance guideline.

Discussion: Some wall coverings have patterns that do not need to be matched.

10–6–18 Observation: Mold and mildew are found on a wall, floor or ceiling surface.

Performance Guideline: Mold and mildew are naturally occurring conditions wherever there is moisture.

Corrective Measure: No corrective action is required by the contractor unless it is the result of construction deficiencies defined elsewhere in this document.

Flooring

Carpeting

11–1–1   **Observation:** Carpet does not meet at the seams.

Performance Guideline: Visible gaps at the seams are considered excessive.

Corrective Measure: It is not unusual for carpet seams to be visible from a standing position. If the carpet was installed by the contractor, the contractor will correct visible gaps at carpet seams.

11–1–2   **Observation:** Carpet is loose or wrinkled.

Performance Guideline: When stretched and secured properly, wall-to-wall carpeting should not unfasten, loosen or separate from the points of attachment.

Corrective Measure: <u>One time only</u> during the warranty period the contractor will restretch or resecure the carpeting as necessary to meet the performance guideline.

Discussion: Consumer is responsible for adhering to the manufacturer's maintenance and cleaning instructions. Excessive moisture during cleaning causes the carpet fibers to stretch and not return to their normal position. This is not a contractor warranty responsibility. The following suggestions help maintain carpeting.

Vacuum regularly and more frequently in high-traffic areas, and everywhere according to a vacuuming schedule.

- Clean spots and spills quickly with products that do not damage the carpet or cause it to resoil quicker.
- Professionally deep clean your carpets every 12 to 18 months to remove embedded dirt and grime.
- Stop dirt at the door by using mats outside and in, taking your shoes off when you enter the house and changing your air filters to reduce airborne dust particles.

11-1-3   **Observation:** Carpet has faded or discolored.

Performance Guideline: Fading or discoloration of carpet is covered by the manufacturer's warranty.

Corrective Measure: No corrective action is required by the contractor.

Discussion: Consumer is responsible for adhering to the manufacturer's maintenance and cleaning instructions. Fading or discoloration may result from the consumer spilling liquids on the carpet, improper cleaning, exposure to sunlight, or from the consumer's failure to properly maintain the carpet.

11-1-4   **Observation:** Carpet appears to be different colors.

Performance Guideline: Carpet for a room should be ordered and installed from a single manufacturer's dye lot. Carpet shade variance is the manufacturer's responsibility.

Corrective Measure: No corrective action is required by the contractor.

Discussion: When viewed under *normal lighting conditions,* carpet may have the appearance of color variations. These differences may result from the direction of the carpet nap or from fibers being crushed on the roll. Over time, vacuuming will make the appearance more uniform.

11-1-5   **Observation:** *Dead spots,* dips or lumps are observed below the carpet surface.

Performance Guideline: Carpeted areas should not have dead spots, or voids dips or lumps that exceed ½ inch.

Corrective Measure: The contractor will repair dips, remove lumps or replace padding in the affected areas to meet the performance guideline.

Discussion: Because carpet padding comprises a number of materials of various densities and feel, there may be an inconsistent feel even with adequate coverage. Some dips may be created with heavy furniture and are not the responsibility of the contractor.

Vinyl and/or Resilient Flooring

11–2–1 **Observation:** Nails or fasteners raised below the flooring surface are readily visible on the surface of vinyl or resilient flooring.

Performance Guideline: Nail or other fasteners from under floor coverings that are raised above the surrounding area and readily visible from a standing position under normal lighting conditions are considered excessive.

Corrective Measure: The contractor will repair or replace flooring.

Discussion: At the contractor's option, the contractor will repair or replace the floor covering in the affected areas with similar materials and in accordance with manufacturer's recommendations. The contractor is not responsible for discontinued patterns or color variations when replacing or repairing the floor covering.

11–2–2 **Observation:** Depressions or ridges are observed in flooring because of subfloor irregularities.

Performance Guideline: Depressions or ridges exceeding ⅛ inch, which are visible from a standing position facing the surface at a distance of 6 feet under normal lighting conditions, are excessive.

Remodeling Specific Guideline: Existing subflooring may have depressions or ridges that exceed the performance guideline. If new floor covering is installed on existing subflooring, the contractor and consumer may agree to disregard the performance guideline to match a pre-existing structural condition.

Corrective Measure: The contractor will take the necessary corrective action to meet the performance guideline. The contractor should not be responsible for discontinued patterns or color variations when replacing or repairing the floor covering. The ridge or depression measurement is taken at the end of a 6-inch straightedge centered over the depression or ridge with 3 inches of the straightedge held tightly to the floor on one side of the affected area. Measure under the straightedge to determine the depth of the depression or height of the ridge.

11–2–3 **Observation:** Vinyl or resilient flooring has lost adhesion.

Performance Guideline: Floor covering should be securely attached to the substrate or underlayment. Some minor voids that exhibit some variance in the sound underfoot may occur. Provided that the flooring material is not otherwise detached and loose at the edges, these variations are not a performance defect.

Corrective Measure: If flooring becomes detached because of improper installation by the contractor, the contractor will repair or replace the affected flooring as necessary. The contractor is not responsible for discontinued patterns or color variations when replacing or repairing the floor covering.

Discussion: The performance guideline does not apply to perimeter-attached vinyl floors.

11–2–4 **Observation:** Seams or shrinkage gaps show at vinyl or resilient flooring joints.

Performance Guideline: Gaps at joints/seams in flooring should not exceed $\frac{1}{32}$ inch in width. Where dissimilar materials abut, the gaps should not exceed $\frac{1}{16}$ inch.

Corrective Measure: The contractor will repair or replace the flooring as necessary to meet the performance guideline. The contractor should not be responsible for discontinued patterns or color variations when repairing or replacing the floor covering.

Discussion: Proper repair can be accomplished by sealing the gap with seam sealer.

11–2–5 **Observation:** Bubbles are observed in vinyl or resilient flooring.

Performance Guideline: Bubbles resulting from trapped air or debris and that protrude higher than $\frac{1}{16}$ inch from the floor are considered excessive.

Corrective Measure: The contractor will repair the floor to meet the performance guideline in accordance with manufacturer's recommendations.

Discussion: The performance guideline does not apply to perimeter-attached floors.

11–2–6   **Observation:** The patterns on vinyl or resilient flooring are misaligned.

Performance Guideline: Patterns at seams between adjoining pieces should be aligned to within ⅛ inch.

Remodeling Specific Guideline: Existing subflooring may have irregularities that result in misalignment. If new floor covering is installed on existing subflooring, the contractor and consumer may agree to disregard the performance guideline to match a preexisting structural condition.

Corrective Measure: The contractor will correct the flooring to meet the performance guideline.

11–2–7   **Observation:** Yellowing is observed on the surface of vinyl or resilient floor covering.

Performance Guideline: The contractor should install vinyl flooring in accordance with the manufacturer's instructions.

Corrective Measure: If the yellowing resulted from improper installation by the contractor, the contractor will repair or replace the flooring. Yellowing resulting from a manufacturer's defect or from the consumer's misuse or lack of maintenance is not covered by the contractor.

Discussion: Some chemical compounds, such as the tar residue from a recently paved asphalt driveway, may cause a chemical reaction with the flooring material and result in permanent damage to the floor. The consumer is responsible for the proper use and maintenance of the floor. Yellowing caused by the consumer's improper use or inadequate maintenance of the floor is not the contractor's or the manufacturer's responsibility.

11–2–8   **Observation:** A resilient floor tile or plank *(LVT, LVP and EVP)* is loose.

Performance Guideline: Resilient floor tiles or planks should be properly installed per manufacturers instructions. Some minor voids that exhibit some variance in the sound underfoot may occur. Provided that the flooring material is not otherwise detached and loose at the edges, these variations are not a performance defect.

Corrective Measure: The contractor will attach loose resilient floor tiles or planks properly per manufacturers instructions. The old adhesive will be removed if necessary to resecure the tiles.

11–2–9   **Observation:** LVP, LVT and EVP flooring makes noise when walking across the floor such as creaks or popping.

Performance Guideline: The contractor should install LVP, LVT and EVP flooring in accordance with the manufacturer's instructions.

> **Remodeling Specific Guideline:** Older subfloors that are unlevel or have existing imperfections may make sounds. Contractor will make best efforts to prepare a level substrate, but it may not be possible and flooring may have some noise.

Corrective Measure: If noises are due to installation, the contractor will repair.

Discussion: If the flooring noise is caused by the subfloor, refer to Guideline 3–3–1 or 3–3–2 for wood subfloors or 2–2–3 for concrete substrate.

11–2–10   **Observation:** The corners or patterns of resilient floor tiles or planks are misaligned.

Performance Guideline: The corners of adjoining resilient floor tiles or planks should be aligned to within ⅛ inch. Misaligned patterns are not covered unless they result from improper orientation of the floor tiles.

> **Remodeling Specific Guidelines:** Existing substrate may have irregularities that result in misalignment. If new floor covering is installed on existing subflooring, the contractor and consumer may agree to disregard the performance guideline to match a preexisting structural condition.

Corrective Measure: The contractor will correct resilient floor tiles or planks with misaligned corners to meet the performance guideline.

Hardwood Flooring

11–3–1   **Observation:** Gaps exist between hardwood floorboards.

Performance Guideline: At the time of substantial completion of the project, gaps between hardwood floorboards should not exceed ⅛ inch in width.

Corrective Measure: The contractor will repair gaps that do not meet the performance guideline.

Discussion: Gaps appearing after installation may be caused by fluctuations in the relative humidity in the home. This is a common seasonal phenomenon in some climates and certain areas of the home that experience significant shifts of humidity. The consumer is responsible for maintaining proper humidity levels in the home.

11–3–2   **Observation:** Hardwood floorboards are cupping or crowning.

Performance Guideline: Cupping or crowning in hardwood floorboards should not exceed $\frac{1}{16}$ inch in height in a 3-inch maximum span measured perpendicular to the long axis of the board. Cupping or crowning appearing after installation are a result of fluctuations in the moisture conditions in the home, causing a noticeable curvature in the face of the floorboards. Cupping or crowning caused by exposure to moisture or humidity fluctuations are beyond the contractor's control and are not the contractor's responsibility.

Corrective Measure: The contractor will correct or repair boards to meet the performance guideline if the cupping or crowning was caused by factors within the contractor's control, only after the moisture content of the flooring and/or the environmental conditions have stabilized.

Discussion: The consumer is responsible for proper maintenance of the floor and for maintaining proper humidity levels and moisture conditions in the home, crawl space or basement.

11–3–3   **Observation:** Excessive lippage is observed along the joints of prefinished wood flooring products.

Performance Guideline: Lippage greater than $\frac{1}{16}$ inch is considered excessive.

Corrective Measure: The contractor will repair lippage in the affected areas to meet the performance guideline if the lippage was caused by elements within the contractor's control.

11–3–4   **Observation:** A wood floor is out of square.

Performance Guideline: The diagonal of a triangle with sides of 12 feet and 16 feet along the edges of the floor should be no more than $\frac{1}{2}$ inch more or less than 20 feet.

> **Remodeling Specific Guideline:** The consumer and the contractor may agree to build a wood floor out of square in order to match or otherwise compensate for pre-existing conditions.

Corrective Measure: The contractor will make the necessary modifications in the most practical manner to any floor that does not comply with the performance guideline for squareness. The modification will produce a satisfactory appearance and may be either structural or cosmetic.

Discussion: Squareness is primarily an aesthetic consideration. Regularly repeated geometric patterns in floor and ceiling coverings show a gradually increasing or decreasing pattern along an out-of-square wall. The performance guideline tolerance of plus or minus ½ inch in the diagonal allows a maximum increasing or decreasing portion of about ⅜ inch in a 12-foot wall of a 12 × 16 foot room.

11–3–5 **Observation:** Voids or skips are observed in the floor finish.

Performance Guideline: Voids that are readily visible from a standing position under normal lighting conditions are considered excessive.

Corrective Measure: The contractor will repair the floor finish in the affected area(s) to meet the performance guideline.

Discussion: This guideline does not apply to distressed, character grade, hand scraped, or other similar flooring finishes.

11–3–6 **Observation:** The top coating on hardwood flooring has peeled or chipped.

Performance Guideline: Field-applied coating should not peel during normal usage. Prefinished coatings are the manufacturer's responsibility.

> **Remodeling Specific Guideline:** Refinishing of existing hardwoods can be affected by previous coatings, including wax, cleaning products and/or pet stains. Consumers recognize that these conditions are not the responsibility of contractor.

Corrective Measure: The contractor will refinish any field-applied finishes that have peeled during the warranty period. Prefinished coatings should not have peeled at the time of substantial completion.

Discussion: The consumer should contact the manufacturer regarding factory-applied finishes that have peeled. It is recommended that painter's tape should never be taped to any hardwood flooring as it may remove the finishes.

11–3–7 **Observation:** Hardwood flooring has buckled.

Performance Guideline: Under normal conditions and usage, hardwood flooring should not buckle.

Corrective Measure: The contractor will repair the affected area to meet the performance guideline if buckling was caused by elements within the contractor's control.

Discussion: Wood floors are naturally susceptible to high levels of moisture. Buckling results from water or high levels of moisture in contact with the floor. Controlling excessive moisture and high humidity levels during cleaning or from other sources is the consumer's responsibility.

11–3–8 **Observation:** Hardwood flooring has released from the substrate.

Performance Guideline: Under normal conditions and usage, hardwood flooring should not lift from the substrate. Some minor voids that exhibit some variance in the sound underfoot may occur. Provided that the flooring material is not otherwise detached and loose at the edges, these variations are not a performance defect.

Corrective Measure: To meet the performance guideline, the contractor will repair the affected area if the lifting was caused by factors within the contractor's control.

11–3–9 **Observation:** Excessive knots and color variations are observed in hardwood flooring.

Performance Guideline: The contractor should install the grade of hardwood specified for the project. All wood should be consistent with the grade or quality specified.

Corrective Measure: The contractor will replace any improper grade or quality of wood.

Discussion: Hardwood flooring is a natural product and consequently can be expected to exhibit variations in color, grain and stain acceptance. This guideline does not apply to distressed, character grade, hand scraped, or other similar flooring finishes.

11–3–10   **Observation:** Slivers or splinters are observed in hardwood flooring.

Performance Guideline: Slivers or splinters should not be visible.

Corrective Measure: The contractor will repair flooring in the affected areas to meet the performance guideline.

Discussion: Slivers or splinters that occur during installation of unfinished wood flooring can be shaved and the area filled prior to sanding and finishing. In most cases, slivers or light splintering in prefinished floors can be corrected. Excessive slivers or splintering of prefinished flooring after installation is covered under the manufacturer's warranty.

11–3–11   **Observation:** Hardwood flooring has visible scratches and dents.

Performance Guideline: At the time of substantial completion of the project, hardwood flooring should not have scratches and dents visible from a standing position under normal lighting conditions.

Corrective Measure: The contractor will repair flooring in the affected areas to meet the performance guideline.

Discussion: The wide varieties of solid and engineered hardwood flooring available to consumers have varying hardness and wear resistance. The contractor is not responsible for the choice of a softer material, which may be more susceptible to damage during or after construction. High-heeled shoes, pets and heavy foot traffic will create scratches and dents in all hardwood floors, some more than others.

11–3–12   **Observation:** *Sticker burn* is observed on the surface of strip flooring.

Performance Guideline: Discoloration from stacking strips on hardwood flooring is considered excessive in certain grades of flooring but is allowable in others.

Corrective Measure: The contractor will repair or replace areas with sticker burn if they are not permitted in the grade of wood specified for the project.

11–3–13   **Observation:** Hardwood flooring is squeaking.

Performance Guideline: Frequent, loud hardwood flooring squeaks are considered deficiencies.

Corrective Measure: The contractor will repair flooring in the affected areas to meet the performance guideline.

Remodeling Specific Guideline: Flooring in an existing home that was not newly installed will squeak and are not covered by this guideline.

Discussion: There are numerous acceptable repairs, such as face nailing, fillers, etc. If the flooring squeaks are caused by the subfloor, refer to Guideline 3–3–1.

Tile, Brick, Marble and Stone Flooring

11–4–1   **Observation:** Tile, brick, marble or stone flooring is broken or loosened.

Performance Guideline: Tile, brick, marble or stone flooring should not be broken or loose.

Corrective Measure: The contractor will replace broken tiles, bricks, marble or stone flooring, and resecure loose tiles, bricks, marble or stone, unless the flooring was damaged by the consumer's actions or negligence. The contractor is not responsible for discontinued patterns or color variations when replacing tile, brick, marble or stone flooring.

11–4–2   **Observation:** Cracks are observed in the tile grout or at the junctures with other materials, such as a bathtub.

Performance Guideline: Cracks in grouting of tile joints commonly result from normal shrinkage conditions. Cracks that result in loose tiles or gaps greater than $1/16$ inch are considered excessive.

Corrective Measure: <u>One time only</u> during the warranty period, the contractor will repair grout to meet the performance guideline. The contractor is not responsible for color variations or discontinued colored grout. The consumer is responsible for re-grouting these joints after the contractor's one-time repair.

Discussion: The use of grout caulk, typically a flexible material, at junctures between tile and other materials can be more effective than grout and is considered an acceptable method of repair.

11–4–3   **Observation:** There is lippage between the transition of marble or ceramic tile floor to another type of flooring surface.

Performance Guideline: Lippage greater than $1/16$ inch is considered excessive, except where the materials are designed with an irregular height, such as handmade tile or large format tiles.

Remodeling Specific Guideline: Because existing subflooring may be uneven and create lippage that exceeds the performance guideline, the contractor and consumer may agree to disregard the performance guideline to match a pre-existing structural condition.

Corrective Measure: The contractor will adjust the affected areas to meet the performance guideline, except where the materials are designed with an irregular height, such as handmade tile or large format tiles.

Discussion: With the increased popularity in irregular and large format tile, lippage can be greater than the performance guideline. Manufacturer's tolerances will supersede this guideline.

11–4–4   **Observation:** A grout or mortar joint is not a uniform color.

Performance Guideline: After the grout or mortar has cured, any color variation that is readily visible from a standing position facing the surface at a distance of 6 feet under normal lighting conditions is considered excessive.

Corrective Measure: One time only during the warranty period, the contractor will repair the joint to meet the performance guideline.

Discussion: Grout or mortar cannot be expected to match exactly in repaired areas.

Miscellaneous

Fireplace and Wood Stove

12–1–1   **Observation:** A fireplace flue/chimney does not consistently draw properly.

Performance Guideline: A properly designed and constructed fireplace and chimney should function correctly.

Remodeling Specific Guideline: In an older remodeled home, energy upgrades as part of the improvements may adversely affect the function of an existing fireplace.

Corrective Measure: <u>One time only</u> during the warranty period, the contractor will repair the flue/chimney, based on the manufacturer's specifications or the design specifications, to draw correctly.

Discussion: High winds can cause temporary negative drafts or downdrafts. Obstructions such as tree branches, steep hillsides, adjoining homes, and certain ventilation systems in the home also may cause negative drafts. Homes that have been constructed to meet stringent energy criteria may need to have a nearby window opened slightly to create an effective draft.

12–1–2   **Observation:** The masonry chimney is separated from the structure.

Performance Guideline: Newly built chimneys will often incur slight amounts of separation. The amount of separation from the main structure should not exceed ½ inch in any 10-foot vertical measurement.

Corrective Measure: The contractor will repair gaps that do not meet the performance guideline.

Discussion: Proper repair can be completed by caulking unless the cause of the separation is a structural failure of the chimney foundation itself. In that case, caulking is not an acceptable repair.

12–1–3   **Observation:** The *firebox* paint is cracked or discolored by a fire in the fireplace.

Performance Guideline: Cracking and discoloration are common occurrences.

Corrective Measure: No corrective action is required by the contractor.

Discussion: The consumer should obtain the recommended paint from the manufacturer if he or she chooses to touch up the interior of the firebox for aesthetic reasons.

12–1–4   **Observation:** A *firebrick* or mortar joint is cracked.

Performance Guideline: Heat and flames from normal fires can cause cracking.

Corrective Measure: Where a firebrick or mortar joint is cracked because of normal fires, no corrective action is required by the contractor.

12–1–5   **Observation:** A simulated firebrick panel is cracked.

Performance Guideline: This is a common condition.

Corrective Measure: No corrective action is required by the contractor.

12–1–6   **Observation:** Rust is observed on the fireplace damper.

Performance Guideline: This is a common condition.

Corrective Measure: No corrective action is required by the contractor.

12–1–7   **Observation:** Water is getting in the home around the chimney.

Performance Guideline: A chimney should be properly flashed and chimney cap properly installed to prevent water from leaking into the house.

Corrective Measure: Contractor will repair to meet the performance guideline.

Concrete Stoops and Steps

12–2–1   **Observation:** Stoops or steps have settled, heaved or separated from the home structure.

Performance Guideline: Concrete stoops and steps should not settle, heave or separate in excess of 1 inch from the home structure.

Corrective Measure: The contractor will use his or her best judgment in making a reasonable and cost-effective effort to meet the performance guideline.

12–2–2   **Observation:** Water remains on stoops or steps after rain has stopped.

Performance Guideline: Water should drain off outdoor stoops and steps. Minor amounts of water can be expected to remain on stoops and steps for up to 24 hours after rain.

Corrective Measure: The contractor will take corrective action to ensure proper drainage of stoops and steps.

Garage

12–3–1   **Observation:** The garage floor slab is cracked.

Performance Guideline: Cracks in a concrete garage floor greater than $\frac{3}{16}$ inch in width or $\frac{3}{16}$ inch in vertical displacement are considered excessive.

Corrective Measure: The contractor will repair cracks in the slab using a material designed to fill cracks in concrete.

Discussion: The repaired area may not match the existing floor in color and texture.

12–3–2   **Observation:** A garage concrete floor has settled, heaved or separated.

Performance Guideline: The garage floor should not settle, heave or separate in excess of 1 inch from the structure.

Corrective Measure: The contractor will use his or her best judgment in making a reasonable and cost-effective effort to meet the performance guideline.

Discussion: The repaired area may not match the existing floor in color and texture.

12–3–3　**Observation:** Garage doors fail to operate properly under normal use.

Performance Guideline: Garage doors should operate as designed.

Corrective Measure: The contractor will correct or adjust garage doors as required, unless the consumer's actions caused the problem.

Discussion: The safety sensors can be easily knocked and misaligned so that the doors will not operate properly. The consumer should avoid storing items near the sensors. Direct sunlight or spider webs can also cause the sensors to indicate that something is blocking the opening and prevent the doors from shutting.

12–3–4　**Observation:** Garage doors allow the entry of snow or water.

Performance Guideline: Garage doors should be installed as recommended by the manufacturer. Some snow or water can be expected to enter under normal conditions.

Corrective Measure: The contractor will adjust or correct the garage doors to meet the manufacturer's installation instructions.

Driveways and Sidewalks

12–4–1　**Observation:** An asphalt driveway has cracked.

Performance Guideline: Longitudinal or transverse cracks greater than ¼ inch in width or vertical displacement are considered excessive.

Corrective Measure: The contractor will repair the affected area to meet the performance guideline using a material designed to fill cracks in asphalt.

12–4–2　**Observation:** Standing water is observed on an asphalt pavement surface.

Performance Guideline: Standing water greater than ⅜ inch in depth should not remain on the surface 24 hours after a rain.

Corrective Measure: The contractor will repair the affected area to meet the performance guideline.

Discussion: Patched asphalt surfaces because of repairs may not match existing surface in color or texture.

12–4–3 **Observation:** The aggregate of asphalt pavement is coming loose in areas other than the edges.

Performance Guideline: Asphalt pavement aggregate should not come loose.

Corrective Measure: The contractor will repair the affected area to meet the performance guideline, using a material designed to repair asphalt surfaces.

Discussion: Patched asphalt surfaces because of repairs may not match existing surface in color or texture. It is not unusual to have aggregate coming loose along the edges of a driveway where the material is not as compactable. Some aggregate under normal circumstances may come loose and is not an indication of a defect in the driveway.

12–4–4 **Observation:** A concrete driveway or sidewalk is cracked.

Performance Guideline: Cracks (outside of control joints) that exceed ¼ inch in width or ¼ inch in vertical displacement are excessive.

Corrective Measure: The contractor will repair affected areas to eliminate cracks that exceed the performance guidelines using a material designed to fill cracks in concrete.

Discussion: Minor concrete cracking is normal and to be expected. Control joints are placed in the concrete to help control cracks and provide a less visible area for them to occur. Cracking can be caused by elements outside of the contractor's control. The repaired area may not match the existing area in color and texture.

12–4–5 **Observation:** Adjoining exterior concrete flatwork sections deviate in height from one section to another.

Performance Guideline: Adjoining concrete sections should not deviate in height by more than ½ inch unless the deviation is intentional at specific locations such as at garage door openings.

Corrective Measure: The contractor will repair deviations to meet the performance guideline.

Discussion: Some areas of the country experience lift or settlement at the junction of the garage floor and the driveway, which occurs because of seasonal fluctuations in moisture and temperature. Repairs will only be made after the effects of the current seasonal fluctuations have subsided and the true determination of repair can be made. The repaired area may not match the existing area in color and texture.

12–4–6 **Observation:** A sidewalk and other exterior concrete flatwork has settled or lifted.

Performance Guideline: Adjoining concrete sections should not deviate in height by more than ½ inch.

Corrective Measure: The contractor will repair deviations to meet the performance guideline.

Discussion: Some areas of the country experience lift or settlement at the junction, which occurs because of seasonal fluctuations in moisture and temperature. Repairs will only be made after the effects of the current seasonal fluctuations have subsided and a true determination of repair can be made. The repaired area may not match the existing area in color and texture.

12–4–7 **Observation:** Water collects or ponds on the sidewalk.

Performance Guideline: Standing water that is ⅜ inch deep on a side-walk 24 hours after the end of a rain is considered excessive.

Corrective Measure: The contractor will repair or replace the affected area to meet the performance guideline.

Discussion: The repaired area may not match the existing area in color and texture.

Wood and Composite Decks

12–5–1 **Observation:** A wood deck is springy or shaky.

Performance Guideline: All structural members in a wood deck should be sized, and fasteners spaced, according to the prevailing building codes and manufacturer's instructions.

Corrective Measure: The contractor will reinforce or modify, as necessary, any wood deck not meeting the performance guideline.

Discussion: Deflection may indicate insufficient stiffness in the lumber or may reflect an aesthetic consideration independent of the strength and safety requirements of the lumber. Structural members are required to meet standards for both stiffness and strength. When a consumer's prefer-ence is made known before construction, the contractor and the consumer may agree upon a higher standard.

12–5–2 **Observation:** The spaces between wood decking board sides are not uniform.

Performance Guideline: At the time of substantial completion of the project, the side-to-side gap between deck boards should not differ in average width by more than $3/16$ inch unless otherwise agreed upon by the consumer and the contractor.

Corrective Measure: If, at the time of substantial completion, the wood decking board sides did not meet the performance guidelines, then the contractor will realign or replace decking boards to meet the performance guideline.

Discussion: The spaces will naturally tend to change over time because of shrinkage and expansion of individual boards. The contractor is only responsible for correct spacing at the time of substantial completion of the project. Replaced decking boards may have a variance in color from board to board.

12–5–3 **Observation:** The spaces between composite decking board sides are not uniform.

Performance Guideline: At the time of substantial completion of the project, the side-to-side gap between deck boards should be installed per the manufacturer's instructions. The side-to-side gaps of individual deck boards should not differ in average width by more than $1/8$ inch unless otherwise agreed upon by the consumer and the contractor.

Corrective Measure: The contractor will realign or replace decking boards to meet the performance guideline.

Discussion: Spacing may be adjusted to keep full-size boards against the edge of the structure or along the edge of the deck. Composite decking must be gapped both end-to-end and width-to-width. Gapping is necessary for drainage and the slight thermal expansion and contraction of composite decking boards. Gapping also allows for shrinkage of the wood joist system. Temperature and humidity variances affect spacing based on the length of the boards installed. Approximately $1/16$ inch per 40-degree temperature change for a 16-foot board can be expected. Replaced decking boards may have a variance in color from board to board.

12–5–4   **Observation:** The end-to-end or butt-to-butt spacing between composite deck boards is excessive.

Performance Guideline: Deck boards should be installed per the manufacturer's installation instructions.

Corrective Measure: Contractor will replace or repair deck boards to meet the performance guideline.

Discussion: Composite decking must be gapped both end-to-end and width-to-width. Gapping is necessary for drainage and the slight thermal expansion and contraction of composite decking boards. Gapping also allows for shrinkage of the wood joist system. Temperature and humidity variances affect spacing based on the length of the boards installed. Approximately 1/16 inch per 40-degree temperature change for a 16-foot board can be expected. Replaced decking boards may have a variance in color from board to board.

12–5–5   **Observation:** The railings on wood decking contain slivers/splinters in exposed areas.

Performance Guideline: Railings on wood decks should not contain slivers/splinters longer than 1/8 inch in exposed areas.

Corrective Measure: At the time of substantial completion, the contractor will repair slivers/splinters on railings. Repair of slivers/splinters after that time is a consumer maintenance responsibility.

Discussion: Slivers/splinters can develop when wood weathers.

12–5–6   **Observation:** A wood framed deck has settled.

Performance Guideline: The plane of the deck surface should not be more than 1/4 inch per foot out of level in 10 feet as measured along the outward bearing points, independent of the attachment to the house.

> **Remodeling Specific Guideline:** The consumer and contractor may agree to intentionally build a wood deck out of level to match or to compensate for inaccuracies in the existing structure.

Corrective Measure: The contractor will repair the deck supports/piers as necessary to meet the performance guideline.

Discussion: A slope of approximately 1/8 inch per foot or more is desirable in the perpendicular direction to the house in order to shed water and prevent ice buildup.

12–5–7   **Observation:** Wood decking boards, railings and/or pickets are split, warped, or cupped.

Performance Guideline: At the time of substantial completion of the project, splits, warps and cups in wood decking boards, railings and/or pickets should not exceed the allowances established by the official grading rules issued by the agency responsible for the lumber species specified for the deck boards.

Corrective Measure: The contractor will replace decking boards, railings and/or pickets as necessary to meet the performance guideline.

Discussion: Pressure-treated lumber is required for exterior applications, and it will likely check, crack and split over time.

12–5–8   **Observation:** A wood deck has applied stain color variations.

Performance Guideline: Stain color variations are not acceptable if they result from improper stain application or failure to mix the stain properly. Stain color variations resulting from other causes — such as weathering or natural variations in the wood used to build the deck — are common and are not covered by this guideline.

Corrective Measure: If the contractor stained the deck, the contractor will re-stain the affected area to meet the performance guideline.

12–5–9   **Observation:** A fastener protrudes from a decking board.

Performance Guideline: Fasteners should not protrude from the floor of the deck.

Corrective Measure: One time only during the warranty period, the contractor will address fasteners that protrude from the floor of the deck so that the heads are flush with the surface.

Discussion: Fasteners should be driven or screwed flush when the deck is installed, but they may pop from a wood deck over time as the wood shrinks and expands.

12–5–10   **Observation:** Fasteners on a wood deck are bleeding.

Performance Guideline: Staining from fasteners is expected on a wood deck.

Corrective Measure: No corrective measure required.

12–5–11   **Observation:** A deck railing lacks rigidity.

Performance Guideline: Deck railings should be attached to structural members in accordance with the prevailing building codes.

Corrective Measure: The contractor will repair deck railings as necessary to meet the performance guideline.

12-5-12   **Observation:** Deck railing lacks rigidity

Performance Guideline: Handrail assemblies and guards shall be able to resist a single concentrated load of 200 pounds, applied in any direction at any point along the top, and have attachment devices and supporting structure to transfer this loading to appropriate structural elements of the building.

Corrective Measure: The contractor will repair deck railings as required to meet the performance guidelines.

12-5-13   **Observation:** Cable railing has *catenary.*

Performance Guideline: Wire/cable rails shall be tensioned sufficiently so that the wire components have no deflection while in a static position. However, care shall be taken so as to not tension the cable so much that the security of the post connection is compromised or the post is deflected. Because cable is flexible, at midspan the cable shall not be able to deflect greater than the 4-inch spacing required by code. Properly tensioned cables may loosen over time due to stretch of the components.

Corrective Measure: For sagging cables, the contractor shall <u>one time only</u> during the warranty period re-tension the cables. For deflection exceeding 4-inch spacing that cannot be corrected by re-tensioning, the contractor will repair as required to meet the performance guidelines.

Landscaping

Note: Remodeling Specific Guideline: *Moving or protecting plants, trees, shrubs and any other landscaping items prior to and during construction are the responsibility of the consumer and must be dealt with before construction begins. Other handling of these items must be specified in the contract to designate the responsible party.*

13–1–1   **Observation:** Tree stumps are left in a *disturbed* area of the property.

Performance Guideline: The contractor is responsible for removing stumps from trees that were on the property in the disturbed area prior to the substantial completion of the project, unless stumps are located within the septic drain field areas or stumps were present prior to the contractor beginning construction.

Corrective Measure: The contractor will remove the stumps from the disturbed area.

13–1–2   **Observation:** Sod, shrubs, plants or trees that were planted as part of the contract are dead.

Performance Guideline: At the time of closing and/or conveyance or occupancy of the property, whichever comes first, any shrub, plant, tree or sod planted by the contractor as part of the contract should be alive.

Remodeling Specific Guideline: All plantings watering and maintenance are the responsibility of the consumer after substantial completion of the project.

Corrective Measure: Any shrub, plant, tree or sod planted by the contractor as part of the contract that is dead at the time of closing and/or conveyance or occupancy of the property, whichever comes first, will be replaced to meet the performance guideline. Replacement material shall be of the same size and variety as the dead or damaged material.

Discussion: After closing and/or conveyance or occupancy of the property, whichever comes first, proper lawn and landscape care are the consumer's responsibility. New landscaping requires adequate watering until roots of plantings have become established. Watering is the consumer's responsibility.

13–1–3 **Observation:** Grass seed does not germinate.

Performance Guideline: Germination is dependent on certain climatic conditions, which are beyond the contractor's control.

Corrective Measure: The contractor is only responsible for seeding. No corrective action is required by the contractor.

Discussion: After closing and/or conveyance or occupancy of the property, whichever comes first, proper lawn care and maintenance are the consumer's responsibility. New landscaping requires adequate watering until roots of plantings have become established. Watering is the consumer's responsibility.

13–1–4 **Observation:** Plants moved by the contractor at the consumer's request or by the consumer during work are dead.

Performance Guideline: Plants that must be moved during the work should be moved, transplanted and maintained by the consumer.

Corrective Measure: No corrective action is required by the contractor.

Discussion: The contractor should not be responsible for delays in the schedule when plants are moved by the consumer. It is the consumer's responsibility to provide a water source to any tree they wish to survive construction.

13–1–5 **Observation:** Existing trees on the property are dead.

Performance Guideline: Even with best efforts, trees in the area of construction activity may not survive because of damage to underground root systems or changes in conditions.

Corrective Measure: No corrective action is required by the contractor.

Appendix

Sample Warranty and Dispute Resolution Contract Language

The following sample comments and clauses are provided for information purposes only, and do not constitute an opinion of law. Builders and remodelers should consult an experienced, local attorney to draft contracts and contractual provisions.

NAHB accepts no responsibility for any inaccuracies or omissions and expressly disclaims any responsibility for damage, liability, loss, or risk, whether personal, financial, or otherwise, that is incurred as a direct or indirect consequence of using, applying, or relying on any of the recommendations and/or information contained in this manual.

Incorporating the Guidelines into a Contract Document

Contract

A contract may validly include provisions of a document that is not physically part of the contract itself — this is referred to as "incorporation by reference." Incorporation by reference is a common tool in the drafting of construction contracts. Matters incorporated into a contract by reference are as much part of the agreement as if they had been set out in the contract word-for-word. Consideration should be given to incorporating the Residential Construction Performance Guidelines by reference into the contract. Below are some examples of how this might be accomplished.

The parties wish to establish a procedure for addressing matters involving the quality or performance of construction. Accordingly, the parties agree that they shall evaluate the work pursuant to the Sixth Edition of the *Residential Construction Performance Guidelines* (RCPG) (Alternative language: the *Residential Construction Performance Guidelines* (RCPG) in effect at the time of the signing of this contract) and that they will abide by the corrective measures dictated by the RCPG in connection with any matters of quality, tolerances, appearances, conditions, construction defects or other aspects of the quality or performance of construction. If an item is not covered in that publication, then local industry custom and practice shall govern.

The Owner acknowledges receipt of a copy of the *Sixth Edition of the Residential Construction Performance Guidelines.*

_____

(Consumer's initials)

Warranty

If a contractor warrants workmanship and materials in a warranty, then the contractor should provide a clear definition of compliance with the terms of the warranty. Specifically, the warranty should specify what constitutes a construction defect or how it will be determined whether an item is covered by the warranty. Without a clear definition, the parties risk having a third party making those determinations for them.

Accordingly, the contract and/or warranty might include a statement such as the following:

> In accordance with the terms of the Warranty described herein, the builder (or remodeler) will repair or replace, at the builder's (or remodeler's) option, any latent defects in the building materials or workmanship that were not apparent or ascertainable at the time of occupancy. What constitutes a latent defect and the appropriate remedy shall be determined in accordance with the guidelines found in the publication *Residential Construction Performance Guidelines, Sixth Edition* (Alternative language: the *Residential Construction Performance Guidelines* (RCPG) in effect at the time of the signing of this contract). If an item is not covered in that publication, then local industry custom and practice shall govern.

> The Owner acknowledges receipt of a copy of the *Sixth Edition of the Residential Construction Performance Guidelines.*

_____

(Consumer's initials)

Quality of Construction Clause

The following clause establishes the *Residential Construction Performance Guidelines* (RCPG) as the referenced determinant for any questions or disputes regarding the quality or condition of construction:

All matters of construction performance shall be in accordance with the criteria contained in the Sixth Edition of the *Residential Construction Performance Guidelines* (RCPG) (Alternative language: the Residential Construction Performance Guidelines (RCPG) in effect at the time of the signing of this contract) Prior to initiating any legal action or alternative dispute resolution proceedings, the parties shall consult the provisions found in the RCPG, and agree to resolve all applicable construction performance questions or disputes in accordance with said provisions. The RCPG shall be binding as to the determination of any issue between the parties involving construction defect, quality, tolerances, appearance or condition in any proceeding brought in arbitration or before a court of law.

The Owner acknowledges receipt of a copy of the *Sixth Edition of the Residential Construction Performance Guidelines.*

_____

(Consumer's initials)

To ensure that the consumer agrees with the specific performance guidelines stated herein, the contractor should review the specific guidelines and the procedures recommended with the consumer before entering into the contract. Reviewing the performance guidelines again at closing or at the walk-through inspection is also recommended.

If there are particular guidelines within this publication that the contractor or consumer does not want to use, they should be specifically excluded in writing from all warranty or contract documents applicable to the project. Likewise, if there are particular issues that are not addressed in the guidelines, then by written agreement the contractor and consumer should refer to those issues in the warranty and/or contract documents.

Mediation and Binding Arbitration Clause

In the event the parties agree to resolve their disputes using arbitration, the parties may wish to specify the following in the arbitration clause:

When determining any matters of quality, tolerances, appearances, conditions, construction defects or other aspects of the quality or performance of construction the arbitrator will employ the *Residential Construction Performance Guidelines for Professional Builders and Remodelers, Sixth Edition* (Alternative language: the *Residential Construction Performance Guidelines for Professional Builders & Remodelers* (RCPG) in effect at the time of the signing of this contract). If a matter is not covered by the RCPG, then local industry custom and practice shall govern.

To ensure that the consumer agrees with the specific performance guidelines stated herein, the contractor should review the specific guidelines and the procedures recommended with the consumer before entering into the contract. Reviewing the performance guidelines again at closing or at the walk-through inspection is also recommended.

If there are particular guidelines within this publication that the contractor or consumer does not want to use, they should be specifically excluded in writing from all warranty or contract documents applicable to the project. Likewise, if there are particular issues that are not addressed in the guidelines, then by written agreement the contractor and consumer should refer to those issues in the warranty and/or contract documents.

_____   _____
(Consumer's initials)   (Contractor's initials)

Glossary

arbitration. A process in which the parties submit their case to a neutral third person or panel of individuals (arbitrators) for a final and binding resolution

arc fault circuit interrupter (AFCI). A specific duplex receptacle or circuit breaker designed to help prevent electrocution or fires by detecting an unintended electrical arc and disconnecting the power before the arc starts a fire

asphalt. A brownish-black solid or semisolid mixture of bitumens used in paving, roofing, damp proofing, and foundation waterproofing.

barn door. A wall mounted door or doors hung by a track and sliding along the face of a wall

beam. A structural member that transversely supports a load

bifold doors. Doors that are hinged at the center and guided by a track

bleed. Discoloration of the materials caused by weathering and/or aging

blocking. Small pieces of wood used to secure, join, or reinforce members, or to fill spaces between members

bow. The longitudinal deflection of a piece of lumber, pipe, or rod

breakline. A dividing point between two or more surfaces

brick veneer. A nonstructural outer covering of brick

buckled. Original plane of material has been altered

bulges. Original plane of material has been altered

bulk moisture. Typically thought of as rain or snow, bulk moisture also includes flowing groundwater

bypass doors. Doors that hang on a track and slide side to side in front of and/or behind one another

cantilever. Outward projection from plane of structure that is supported at one end

catenary. The curve a hanging flexible wire or chain assumes when supported at its ends and acted upon by a uniform gravitational force

caulking. Material used to fill a void between two other materials

cement board. A cementitious manufactured building material

checking. Cracks in wood

chimney cap. A metal or masonry surface that covers the top portion of a chimney and prevents the penetration of water

circuit. The complete path of electricity away from and back to its source

circuit breaker. A device that automatically interrupts an electrical circuit when it becomes overloaded or is used to turn off a circuit

closed crawl. Crawl space with no outside ventilation

closing. The final phase of mortgage loan processing in which the property title passes from the contractor to the consumer

cold joint. A joint in poured concrete where the pour terminated and continued

concrete flatwork. Horizontal poured concrete surface

condensation. The conversion of moisture in the air to water

condition[ed] crawl. Crawl space with mechanical ventilation and/or heat/cool source

concrete control joint. Grooves and/or saw cut manually made on concrete to help control where the concrete may crack

coped. A piece of trim material that is cut to fit together in profile to another piece of trim

corner bead. A strip of metal, plastic, or vinyl used to shape corners before finishing a wall and fiberglass

crawl space. An area under a home that is not a basement or cellar and is not considered habitable space

crowning. A condition occurring when the center of a board is higher than its outside edges

cupped. A condition wherein the center of a board is lower than its outside edges

damper. A device used to regulate airflow

dead spot. Areas below a carpeted surface where padding seems to be missing or improperly installed

deflection. The bend of a joist, truss or beam under a load

delamination. Split or separation of a laminated product into layers

dew point. The temperature at which moisture in the air condenses into liquid

dimensional lumber. Building material of stated length. Lumber that is cut to predetermined sizes.

disturbed area. Any area adjacent to a dwelling where original vegetation has been altered or removed

doorjamb. Vertical piece of door frame to support a lintel

doorknob, deadbolt, lockset. Hardware to secure, allow a door to open/close

downspout. A pipe that carries rainwater from the roof to the ground or to a groundwater management system

drywall. Gypsum board

duct. A round, oval or rectangular pipe used to transmit and distribute warm or cool air from a central heating or cooling unit, or a pipe connected to a bath or kitchen exhaust fan to transmit air to the exterior of the home

ductwork. A system of ducts, dampers, plenums, and fans that creates a continuous passageway for the transmission of air

eave. The lower or outer edge of a roof that projects over the side walls of a structure

efflorescence. White powder that appears on the surface of masonry walls. It is usually caused by moisture reacting with the soluble salts in concrete and forming harmless carbonate compounds.

EVP. Engineered Vinyl Plank is a vinyl flooring that has a realistic hardwood look (and feel) and is durable, waterproof and has a strong high density fiberboard core.

fasteners. A product to attach, fix, or join

finish flooring. The top flooring material that covers the subflooring surface, such as carpeting, hardwood, tile, laminate or vinyl

firebox. An enclosure for a fire in a fireplace

firebrick. A brick that can withstand very high temperatures that is used in a fireplace

flashing. Strips of metal or plastic used to prevent moisture from entering roofs, walls, windows, doors, and foundations

floating floors. A floor that does not need to be nailed or glued to the floor underneath it

floor joist. A horizontal framing member to which flooring is attached

footing. The system at the base of a foundation wall that supports and distributes loads from the foundation to the ground

foundation. That part of a building which starts below the surface of the ground and upon which the superstructure rests

gable end. Triangle wall section at the end of a pitched roof

gable vents. Vents to allow air flow at gable ends

grid, grille, and muntin. Strips of wood, metal, or plastic installed within two pieces of glass or on the inside and/or exterior surface of the glass that divide a window into panes

ground fault circuit interrupter (GFCI). A type of circuit breaker that is extremely sensitive to moisture and changes in resistance to an electrical current flow. A GFCI protects against electrical shock or damage.

gypsum. Hydrous calcium sulphate mineral rock used to make wallboards

gypsum wallboard. A type of drywall or sheetrock

hammering. See water hammer

hardwood. A term used to designate wood that is from deciduous trees, which lose their leaves annually

header. A structural member placed across the top of an opening to support loads above

hip. The external angle formed by the upper juncture of two sloping sides of a roof

honeycomb. Pits, surface voids, and similar imperfections caused by air entrapped at the concrete and concrete form interface

HVAC. Heating, ventilating, and air conditioning

I-joist. An "I" shaped engineered wood structural member

jamb. The side framing or finish material of a window, door, or other opening

joist. An on-edge horizontal lumber member, such as a 2 × 6, 2 × 8, 2 × 10, 2 × 12, I-beam, truss, or other material which spans from wall to wall or beam to provide main support for flooring, ceiling, or roofing systems

joist hangers. Metal straps or brackets used to connect framing members

kick plates. Protective cover at bottom of door panel

knots. A hard node of the tree as it appears on cut lumber

LVT. Luxury Vinyl Tiles. A high-performance semi-flexible floor covering composed of several laminated layers which mimics tile and is waterproof

LVP. Luxury Vinyl Planks. Vinyl flooring that is made up of planks, as opposed to one large sheet of vinyl which mimic hardwoods and is waterproof

lath. Any material used as a base for plastering or stucco surfacing

lippage. The difference in surface alignment between two materials, such as tile or stone slabs

louver. An opening with horizontal slats that allows for the passage of air, but not rain, light, or vision

lower chord. Lower design member of a truss system

manufacturer's warranty. The warranty provided by a manufacturer on a specific product

masonry. Brick, stone, concrete block, and other similar building materials bonded together with mortar

mediation. A process whereby the parties meet voluntarily to negotiate a private and mutually satisfactory agreement aided by a neutral third party

membrane roofing. A type of roofing system for buildings with flat or nearly flat roofs designed to prevent leaks and move water off the roof

mitered. Two pieces of trim beveled at 45-degree angles to form (when joined) a 90-degree corner. It also applies to any joint between trim materials that is cut at ½ the angle of their intersection. It is the opposite of a coped inside corner.

mortar. An adhesive and leveling material used in brickwork, stone, block, and similar masonry construction. Also, to set exterior and interior tile.

nail pop. The protrusion of a nail or screw in a panel of drywall or underlayment usually attributed to the shrinkage of or curing of wood framing

natural stone. For the purposes of this manual, any of the following:
• Igneous rocks such as granite
• Sedimentary stone such as limestone, shale, sandstone, onyx, and travertine
• Metamorphic rock – such as marble and slate
Typical residential uses of natural stone include wall and floor surfaces, counter tops, stairs, and other decorative and structural elements

normal lighting. Moderate light level, as measured in lumens, associated with interior conditions on an overcast day. A lumen rating between 150 and 250 is considered normal. Light should not be obstructed with curtains, blinds or other window or door coverings

occupancy. The time at which the consumer possesses or resides in the house

parging. A rough coat of mortar applied over a masonry wall

picket. A vertical member in a railing system used to create a fall barrier

pier. Vertical supporting structure

pitch. The degree of incline in a sloped roof or structure

pitting. Small cavities in a concrete surface

plane. A surface in which any two points are chosen, a straight line joining them lies wholly in that surface

plumb. A measurement of true vertical

pocket door. A door that slides into a wall usually into a pocket-style hardware system

ponds. Accumulation of water before draining/absorption/evaporation occurs

radiant floor. A floor that is heated by a hot water system with pipes or by electrical mats, or cables that are placed in the floor

rafter. Structural members that shape and form the support for the roof deck and the roof covering

register. A louvered device that allows air travel from the ducts into a room

reveal. The space between two adjacent components

ridge. The horizontal line at the junction of the top edges of two sloping roof surfaces

riser (stairway). A vertical stair member that supports a tread

roof ridge. The apex of a roof system

roof sheathing. Boards or sheet materials nailed to the top edges of trusses or rafters to tie a roof together and support the roofing material

rust marks. Stains caused by the oxidation of metallic components

scaling. Peeling or flaking of surface of concrete

setting. The driving of a fastener flush or below the surface of a material

settlement. The act of soil compacting due to natural or artificial pressure

shading/shadowing. A slight variation/difference in color

shakes. Split wooden shingles that are random in thickness

sheathing. The application of panels to the face of framing members. Also known as "decking."

shim. A thin, tapered piece of material (usually wood) that is used to adjust or provide support or alignment.

shrinkage cracks. Surface cracks due to curing process

siding. Building material used to protect/decorate the side of a structure

skip. A natural depression below the surface of a planed board

skirt. In a stair system, the board that runs along the ends of each step. On masonry it is the outer layer of protection in the system to cover step or other flashing pieces.

slab. A concrete floor or surface

soffit. The enclosed undersurface of an eave which may be vented or non-vented

spalling. The breaking away of a small piece of concrete

sticker burn. The discoloration from stacking strips which occurs during the drying and storage of hardwood boards

stucco. An exterior finish product composed of sand, lime, and cement installed over a concrete wall or lath system

subfloor. A floor decking material installed on top of the floor joists over which a finish floor is to be laid often with another layer of sheathing for support and an appropriately smooth surface for some flooring types

substantial completion of the project. The point in a project when areas of the residence are functional for their intended use as defined by the contract

sump pump. A pump installed in a crawl space, basement, or other low area to discharge water that might collect.

swale. A shallow depression in the ground that is used to drain water

tannin. A yellowish or brownish organic substance present in some woods

tarnish. To dull or lose luster from exposure to natural conditions

telegraphing. A condition of a subsurface projecting through the finish material, as with existing shingles through a new layer of shingles.

tread. A horizontal stair member. The surface one steps upon when walking up or down stairs.

trowel marks. Impressions in dried joint compound made by a trowel or other drywall finishing tool

truss. An engineered assembly of wood or metal components that is generally used to support roofs or floors

tuck pointing. Application of building material to fill voids

underlayment. A building material used to provide a layer of protection

upper chord. Upper design member of a truss system. Opposite of a lower chord.

vapor barrier. Material to prevent moisture infiltration

vented crawl space. Crawl space vented by air infiltration

vertical displacement. Movement of a building material in a vertical direction

warp. To distort out of shape

warranty period. The duration of the applicable warranty agreed upon by the contractor and the consumer in a contract

water hammer. A hammering or stuttering sound in a pipeline that sometimes accompanies a sudden and significant change in the flow rate of the fluid through the pipeline

weather stripping. Material placed around doors, windows, and other openings to prevent the infiltration of air, dust, rain, or other elements

weep hole. A small hole in a wall or windowsill that allows water to drain

Resources

Builderbooks

The following resources are available at BuilderBooks.com to help build your business:

NAHB Business Management & Information Technology Committee. *The Cost of Doing Business Study, 2022 Edition,* Washington, DC: BuilderBooks, 2022.

Van Note, Steve. 2021 *Home Builders' Jobsite Codes: A Quick Guide to the 2021 International Residential Code,* Washington, DC: BuilderBooks, 2022.

Elkman, Mollie. *The House That She Built,* Washington, DC: BuilderBooks, 2021.

Shinn, Emma. *Accounting and Financial Management for Residential Construction, Sixth Edition,* Washington, DC: BuilderBooks, 2021.

National Association of Home Builders and the International Codes Council. *ICC 700-2020 National Green Building Standard®,* Washington, DC: BuilderBooks, 2021.

Dixon, Dennis. *Finding Hidden Profits: A Guide for Custom Builders, Remodelers, and Architects,* Washington, DC: BuilderBooks, 2017.

NAHB Contracts

NAHB offers a variety of residential construction contracts that are affordable, convenient and reliable, exclusively for home builders and remodelers. Each contract product is delivered as a downloadable Word document. You will save hours of work—which means saving hundreds of dollars—with every contract. Choose the product and pricing option(s) that best meet your needs. **Visit nahbcontracts.com.**

BizTools

Visit nahb.org/biztools and log on as an NAHB member for a variety of business management resources to help you work more profitably and productively. Articles and resource materials are organized by category, including accounting and financial management, business and strategic planning, customer service, construction management, human resources, sales and marketing, and home technology solutions.

Index

A

air conditioning, 82–83
air handler vibration, 82.
    See also heating and cooling
air surface voids, 10
appliances, electrical outlets for, 74
arbitration, 135
arc fault circuit interrupter (AFCI), 71
asphalt, 122–123
attic ventilation, 51–52

B

backsplashes, 99
balconies, 91
banging sounds, 63
barn doors, 88
basements, 8–12
bathtubs, 64–65. *See also* plumbing
beams, 17–18, 24
bifold doors, 87
binding arbitration, 135
blisters, 102–103
bowing
    basement walls, 9–11
    ceilings, 23, 49
    columns, 14–15, 18
    exterior trim, 46
    framing, 23
    roofs, 49–50
    siding, 35, 37–38
bricks, 41–43, 117–118
brush marks, 105
bubbling, 110
buckling, 36, 54, 115
bypass doors, 87

C

cabinets, 94–98
cable railings, 128
cable television wiring, 76
can lights, 76
carbon monoxide detectors, 75
carpet, 107–108
caulk
    bathrooms, 64
    exterior doors, 30–31
    exterior trim, 46
    exterior walls, 25
    siding, 35, 38–39, 41
    windows, 24
ceiling, bowing, 23, 49
ceiling fans, 75
ceilings
    bowing, 23, 49
    paint, 101–102
    stains on, 37
    textured, 104
central vacuums, 78
checking, 24
chimneys, 57, 119–120
chips
    concrete floors, 8
    countertops, 98, 100–101
    fiber cement board siding, 40–41
    hardwood flooring, 114
    plumbing fixtures, 64
circuit breakers, 71–73
climate control
    air infiltration and drafts, 77–78
    ducts and airflow, 79–80
    heating and cooling, 80–83
    humidity and condensation, 78
    ventilation, 84–85
closed crawls, 14
cold joints, 11

D

H

hammer marks, 93
handrails. *See* railings
hardwood, 112–117
heating and cooling, 80–83
honeycomb, 10
humidity. *See also* condensation;
     moisture
  cabinets, 95
  climate control, 78, 81
  exterior doors, 28–34, 77
  flooring, 19–20, 113
  membrane roofing, 56
  sewer odor, 68
  siding, 35–36
  trim affected by, 92–93
  windows, 27–28
hurricanes, 52
HVAC, 80–83

I

ice buildup, 52, 78
incorporation by reference, xvi, 132
insulation, 25, 61, 77–78
interior finish
  cabinets, 94–97
  countertops, 97–101
  doors, 87–90
  stairs, 90–91
  trim and moldings, 92–93
  walls, 101–106

J

jambs, 88
joints, 11, 103

L

laminate countertops, 97–98, 100
land, settling of. *See* settlement
landscaping, 1–3, 129–130
lap marks, 105
latches, 90
latching, 29–30
lath and plaster, 101
leaks. *See* moisture; water
levelness
  cabinets, 94, 96
  countertops, 98
  decks, 126
  drywall and, 102–103
  foundations, 5–7
  siding, 36, 39
  subfloors, 19–21, 112
  windows, 27
light tubes, 59
lighting, 74–76
locksets, 32–33, 90
louvers, 51
low-voltage transformers, 75
lumps in carpet, 108

M

manufacturers, scope of responsibilities,
     xiii–xiv
manufacturer's warranty, xv
marble, 117–118
masonry, 15, 41–42. *See also* chimneys
measurements, taking with coins, xvii
mediation, 135
mildew, 106
moisture. *See also* condensation;
     humidity; water
  barriers and flashing, 24–25
  basement walls and floor, 12
  carpets and, 107
  concrete flatwork and, 123–124
  crawl spaces, 13–14

Y